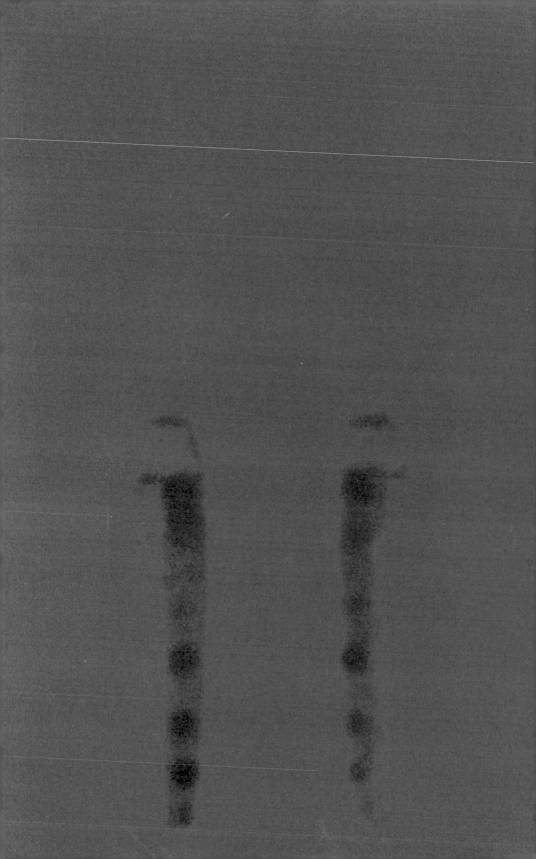

UNDERWATER ENGINEERING

Underwater Engineering

RON GOODFELLOW

Cap. be

Petroleum Publishing Company
Tulsa, Oklahoma

Library of Congress Catalog Card Number: 75–21194
International Standard Book Number: 0–87814–065–4
Printed in U.S.A.

Contents

	Preface	xi
1	The Underwater Engineer	3
2	Diving Technology	14
3	Inspection and Maintenance	33
4	Underwater Vehicles	56
5	Underwater Power Sources	87
6	Subsea Oil Production	103
7	Communication and Navigation	140

Acknowledgements

IN the preparation of this book, numerous companies and individuals have assisted and I would like to acknowledge with sincere thanks the following organizations (and apologies in advance for any omissions):

Abu Dhabi Marine Areas Ltd. (ADMA)
Admiralty Experimental Diving Unit
B. S. & B.
British Petroleum Co. Ltd.
Comex Diving Ltd.
Comex-John Brown Ltd.
Comex S. A.
Cooke Brothers
C. J. B. Offshore Ltd.
D. H. B. Ltd.
Diving Unlimited
Drägerwerk
Exxon
Fishing News International
General Electric Company
Graseby Instruments
Grumman Aerospace Corp.
Hitachi Ltd.
Institute of Oceanographic Sciences
International Hydrodynamics Co. Ltd. (Hyco)
Intersub
Japan Society for the Promotion of Machine Industry
Japan Ships Machinery Development Associates
Jet Research Center Inc.
Kamatsu Ltd.
Kawasaki Heavy Industries
Lindquist S. A.
Lloyds Register of Shipping

Lockheed Petroleum Services
Marine Unit Technology
Mitsubishi Development Corporation
C-E Natco
Nordic Offshore Drilling Co.
Northern Offshore
Novosti Press Agency
Oceaneering International
Ocean Systems
Offshore Rig Data Services
Perry Submersibles
Reynolds Metals
Royal Naval Physiological Laboratories, Alverstoke
Royal Navy
SEAL
Shell Oil Co.
Siebe Gorman
Simrod A. S.
Sonardyne
Sub Ocean Services
Sub Sea Oil Services SpA
Sumito Shipbuilding and Machinery Ltd.
UMEL
U.S. Navy
Vickers Oceanics Ltd.
White Fish Authority
Woods Hole Oceanographic Institute

To the following persons, I would like to offer special thanks. On Chapter 1 my thanks go to Dr. Tom Gaskell for his assistance on oceanographic matters and his continued support and encouragement to persevere with the book. To John Bevan (CxJB's Senior Diving Technologist) who ably assisted in the re-draft of Chapter 2. To my colleagues Nick Cresswell (Naval Architect) and Tony Kidman (Offshore Systems Engineer) for their assistance with Chapters 3 and 4. To Kenji Okamura (Managing Director, Mitsubishi Development Corp.; President, E.C.O.R.) for his help in obtaining photographic material for Chapter 4. To my diving buddy

Alan Webb (BP) for his contribution to Chapter 5 and Ken Haigh (Officer in Charge, Admiralty Experimental Diving Unit; Vice President, Society for Underwater Technology) for his assistance on underwater power sources. My thanks to Graham Mead (Study Manager (Subsea), CJB Offshore) for his constructive comments on Chapter 6 and Ron Geer (Shell) for his help in providing photographic material. And to Gregory Haines and John Partridge for assistance and material in Chapter 7.

Many more friends and colleagues than could possibly be mentioned here have generously provided data and illustrations for this book and where possible this has been acknowledged in the text.

Finally, I acknowledge the continued assistance of my secretary, Peggy Day, who is able to transcribe illegible English into passable English and Madeleine Siddiqui for proof reading the book.

Preface

THE offshore operations in European concessional waters in the last decade have required new techniques and innovations.

The use of concrete for gravity structures, submersibles for pipeline inspection, and the advances in pipelaying techniques have made considerable demands on engineers. Not only the offshore oil and gas industry, but the fishing industry also requires the skills of engineers with an understanding of this environment.

The nations of the world are turning to the oceans for their ever-increasing demands for food and energy, and this makes the next decade an exciting period for those involved.

We shall see divers progress beyond 1,200 ft—once thought to be the depth of the "helium barrier"—developments in submersibles and TV systems, making efficient and clear records of the seabed and structures upon it, and improvements in manipulators and robots for the numerous tasks that are required with subsea production systems.

The last decade has seen a tremendous advance in this field, and the next decade should see continued growth with improvements of techniques and skills. There is still much to be done. In particular, effective systems for inspection and improved or new techniques for maintenance will be developed. The uses of manned and unmanned vehicles will increase and their applications are being more clearly defined.

Within this vast domain, the ingenuity and skills of man will continue to play the leading role. His designs, equipment, fabrications, and installations will be vital to all, but it is important that the underwater engineer understands fully his role and responsibility to the benefit of mankind, and that he progresses safely and economically in this very demanding field.

UNDERWATER ENGINEERING

1

The Underwater Engineer

TO work on or under the seas or on the sea bed has been and will continue to be a challenge to the engineering ability of man.

Now, more than ever, man is turning to this vast domain for the vital resources of food and energy that he so badly needs. The riches of the seas are coming within present engineering attainment, but the full potential has yet to be realized. The skills of the mechanical, electrical, structural, and instrumentation engineers combine together to challenge the common factor—the sea.

The engineering activities are wide and varied and range from transportation and fishing to offshore exploration and diving. The problems are international and multi-disciplinary, and the triangle of technical, economic and political aspects is always there.

The engineer becoming involved in this activity has four main lessons to learn:

1. Never underestimate the sea, the power and strength of its winds, waves, tides, and currents. These harness a tremendous amount of energy and impose considerable forces on structures.

2. Come to terms with the environment. The engineer should be trained to dive, as this skill would be invaluable. Perhaps even salt water in the veins would help.

3. Design for the environment and take into account the advantages of pressure, buoyancy, and a controlled environment at depth, with regard to temperature. Remember: Don't make things complicated. Finding faults under water can be time-

3

consuming, generally taking three times as long as on the surface, but costing five times as much.

4. When carrying out development work on equipment and techniques for the underwater environment you need time. However, more thought and time should be spent on the surface and then in shallow water, and only then should deep water and exposed conditions be contemplated. Long-term reliability and durability cannot be determined in days.

The time factor is extremely important, as experience cannot be bought. Therefore, this is an investment rather than a purchase. Many more nations are now investing time and money in the development of the oceans. Countries like the U.S.A., France, and Norway are notably taking the lead from those nations, such as the U.K., who have had a long traditional involvement with the seas.

Many of the offshore innovations are being tried in European waters, and now technology is being further advanced due to the energy crisis. Sources of energy beneath the sea, which in the past were uneconomic to exploit, now require the engineering skills necessary for their extraction.

The chapters which follow deal with separate aspects of the field of underwater engineering, but it must be clearly understood that this area is progressing rapidly, and many engineering skills and disciplines are involved in any one project. The next ten years will see diving progressing to depths beyond 300 meters, more use of underwater vehicles and manipulators, and sea-bed well-heads producing oil from greater depths and more exposed locations.

The Environment

The oceans cover 70.8% of the earth's surface, and this marine area totals 139,480,000 square miles. A study of a bathymetric chart of the world's oceans will show that most of the continents are entirely surrounded by a platform covered by shallow water. This is known as the continental shelf.

Of the total marine area of the world, 7½% comes within the

4

continental shelf at a depth generally not exceeding 200 meters (600 ft). The width of the shelf varies from almost zero, as along parts of the west coast of Africa, to a maximum of 1500 km from the coasts of Louisiana and Texas.

The significance of the shelf lies in the fact that it represents the seaward extension of the continents which were submerged, following the melting of the ice, after the Ice Age. This means that the bed rock on the shelf actually forms a continuation of the geology of the adjoining land area.

The surface of the shelf may be smooth and consist of bed rock but more often it is irregular, coated with recent sediments and cut by submarine valleys or canyons. In places these line up with major rivers on land, such as the Congo River of West Africa, or the Hudson River on the eastern seaboard of the U.S.A.

These canyons have been cut by torrents of mud and sand-laiden water which have travelled at high speeds down the continental shelf and then discharged their load when the water was checked on the sea floor beyond the edge of the shelf. The sediments deported in this way are known as turbidites.

Beyond the continental shelf the sea floor consists of the continental slope and the abyssal plain. Away from the edge of the continental shelf, the sea floor dips more strongly with a gradient of about 1 in 15 and falls steadily down the continental slope to the abyssal plain where water depths average about 6000 meters (20,000 ft).

The plain is dissected by deep trenches such as the Marianas Trench in the Southern Pacific where the deepest part of the world's oceans has been recorded at a depth of 10,907 meters (35,640 ft) by the Challenger Expedition of 1951.

The deep ocean. These deeps and parts of the abyssal plain contain deposits made up of clays and animal and plant remains which have accumulated very slowly over long periods of time. Cores taken from these deposits show a record of deposition extending over tens of thousands of years.

The surface of the abyssal plain is also interrupted by submarine ridges as well as by submarine volcanoes, some of which are active as in the South Pacific. Others are long since dead.

The submarine ridges are of profound importance in understanding the origin of the continents and oceans. They have been recognized in broad outline for many years, but their details have only recently been studied. It is now seen that there is an almost continuous submarine mid-ocean ridge extending for some 40,000 miles through the oceans of the world.

The ridge is broken by a fracture or faults, and the crest itself is rifted by deep valleys. The best known of these ridges is the Mid Atlantic Ridge. This extends southwards from near Iceland, to form the North Atlantic Ridge, and then extends further south to the South Atlantic Ridge midway between the continents of South America and Africa.

Below the veneer of sediment on the deep ocean floor is basaltic rock which overlies the deep mantle of the earth's crust. This contrasts with the continental areas which, together with the surrounding continental shelf, are floored with granitic rocks. These in turn overlie the basaltic substratum.

The measurement of temperature at different depths in the sea, together with a scientific following of the sea water by means of the salt content, has led to a knowledge of currents which exist at depths, as well as at the surface. Such observations have led to a theory of general circulation of water in oceans. Cold water sinks mainly in high latitudes, flows toward the equator, and is replaced by warmer water flowing polewards from the surface. Much of our knowledge is still built up from studies of the distribution of water temperatures, salinity, and density.

Cold-water sources. There appear to be two main sources of cold deep water. One is in the Weddell Sea on the border of Antarctica, which produces very cold water found at the bottom of the Atlantic, Indian and Pacific oceans. The other, slightly warmer and more saline, sinks in the extreme north of the Atlantic Ocean. The water spreads southwards above the bottom water, and in the Antarctic it appears as a relatively warm intermediate layer between the bottom water and colder but fresher surface water.

The most obvious effect of the wind is to make waves, but part of the energy transferred to water drags it along as a current. The forces are, however, small compared with the weight of water, so that it must take some time to build up a current. Studies in the Indian Ocean have shown that it takes a month or so for the S.W. Monsoon to set up the powerful Somali Current along the east coast of Africa, north of the Equator.

Sometimes the pattern of winds can set up slopes in the sea surface, piling water up against the land. The horizontal pressure gradients produced in this way provide currents, in the depth of the ocean as well as at the surface. The Gulf Stream seems to be largely such a current, maintained by the northeast trade winds and the equatorial current.

The earth's rotation also has an important effect on tides and currents even in small seas. It is least at the Equator and increases towards the Poles. It is only recently that theories and observations have been made concerning the probable combined effects of wind, the earth's rotation and density differences.

Perhaps the most important consequence of ocean currents is their effect on climate. By carrying large volumes of hot and cold water they influence the pattern of energy transfer from the ocean to the atmosphere. However, as yet too little is known of the complex interactions of ocean and atmosphere.

Current Activities

It is of course fish which comprise the bulk of the food we get from the sea; present catches amount to some 40 million tons a year, Table 1-1. The reason lies in the convenience with which fish can be caught and that they present an immediately acceptable product.

Although the fish represent a comparatively small part of the organic matter produced in the sea, fish have concentrated this material which, even though more abundant, would have been difficult to catch and process. The occurrence of fish in shoals has further assisted in harvesting the food.

To maintain a fishery at its most favorable level some means of control has to be employed but the methods used must be designed to prevent overfishing and yet not close the industry down economically. Currently, methods that are used are closing of cer-

TABLE 1-1
The Catches of Fishes of Different Groups of Oceanic Areas (thousands of tons)

	Flounders, halibut, soles	Cods, hakes, haddocks	Redfishes, basses, congers	Jack mackerels, mullets, sauries	Herrings, sardines, anchovies	Tunas, bonitos, billfish	Mackerels, snoeks, cutlassfish	sharks, rays, ratfish
North-west Atlantic	298	1524	353	14	859	18	398	39
North-east Atlantic	366	3964	691	295	1889	53	378	103
West central Atlantic			201	47	635	46	2	8
East central Atlantic	13	49	294	549	957	209	300	17
Mediterranean and Black Sea	11	32	102	81	431	37	13	14
South-west Atlantic	7	134	177	68	184	12	9	24
South-east Atlantic	3	1121	88	375	1102	22	72	6
West Indian Ocean	10	7	324	69	331	136	149	116
East Indian Ocean	1		89	26	58	49	33	29
North-west Pacific	310	78	783	666	961	336	1558	54
North-east Pacific	257	1623	307		114	20		3
West central Pacific	6		265	422	158	186	198	30
East central Pacific	6		29	27	222	317	1	8
South-west Pacific	1	37	49	31		82	19	5
South-east Pacific	1	99	25	198	5638	71	9	10
Total	1228	11472	3774	2870	13535	1594	3131	466

tain ground for limited parts of the year, fixing size limits below which it is illegal to land fish, and regulating the mesh size of nets.

New fishing grounds offer possibilities of expansion in several parts of the world. (See Fig. 1-1.) If the present annual rates of growth of fisheries can be maintained, it may be possible to increase the harvest of fish and other animals from the ocean about fourfold, to, say, 160 million tons per year at the beginning of next century.

Food production aside, it should not be forgotten that marine organisms have other products of use to man. Although the sea contains the vast majority of the known elements they are for the most part at such low concentrations in the water that their extraction is not economically feasible.

However, marine organisms do concentrate many of these elements, and subsequent extraction from the organism becomes a much more feasible proposition. Iodine, for example, has for a very long time been extracted from the ash of burnt seaweed.

Man has harvested salt from the sea by solar evaporation since the dawn of civilization. By far the greater part of the salt produced in the world today is mined from deposits laid down under-

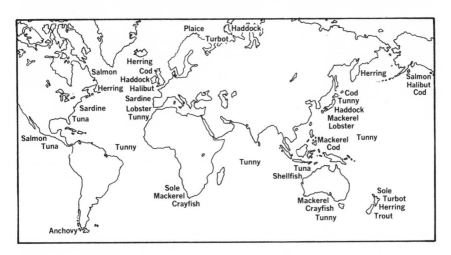

Fig. 1-1. *World's main fishing grounds, and types of fish at various locations.*

9

ground by the evaporation of ancient seas. There are, however, areas of the world where no salt deposits exist and sea water evaporated by solar means is still a very important commercial industry, as is the production of fresh water. However, it is only in the last forty years that serious exploration of other chemicals has been considered; namely, bromides, magnesia, and to some extent potassium.

Serious studies have also been made of the possibility of extracting uranium from sea water.

Natural resources. For many years industry has been concerned with the winning of natural resources from the sea bed and from the rocks which lie below. Natural gas and crude oil as well as coal have been extracted from below the continental shelf for many years, but it is only within the last two decades that the potential wealth of the sea bed has come to be realized.

Useful mineral sources vary from the deposits of sand and gravel on the surface of the sea bed, which can be won by dredging or similar techniques, to those of the crude oils requiring drilling techniques to depths of 15,000 ft below the sea bed.

Coal has been exploited by undersea mining off the Scottish coast for several hundred years, as have Cornish tin and iron ore and other ores elsewhere in the world.

Natural gas and crude oil are the most important of all minerals. Currently some 17% of the world's crude and 4% of natural gas are being produced from continental shelf fields. These are mostly in sedimentary forms which extend onshore. By the 1980's this will have increased to 38%.

Less common and not so well known are the occurrences of nodules of phosphate on the upper parts of the continental slopes. These occur in local concentrations, such as off the coast of Southern California, where reserves are estimated at 1.5 billion tons. They have not yet been commercially exploited, but they do represent a strategic reserve of phosphate.

In the deeper parts of the ocean on the abyssal floor, are found nodules of another mineral, manganese oxide. They contain a high percentage of manganese about 24% and the manganese oxide is often combined with other minerals.

Quite a new source of potential economic interest has come to light in recent years by the discovery in 1964 to 1966 of highly concentrated brines in the Red Sea. These contain very high contents of metals such as iron, manganese, zinc, and lead.

Offshore Oil and Gas

Offshore oil production rates worldwide are running at about 10×10^6 bpd, with a forecast rise to 40×10^6 bpd by 1980. The UK sector of the North Sea is estimated to have a production capability of $3\text{-}4 \times 10^6$ bpd by 1980 (compared with current UK consumption in the order of 2×10^6 bpd).

The large forecast rise in offshore activity indicates the proposed extent of man's activity in and below the sea over the next decade. See Fig. 1-2. While many other mineral deposits are located both in and under the sea, and notwithstanding the potential for such activities as fish farming, it is oil and gas production that will drive man under the sea in the coming years.

Current offshore activity requires man's subsea intervention for a variety of reasons:
- Pipeline, flow-line and loading-buoy hose connections.
- Blow-Out Preventor (BOP) stack installation when drilling.
- Inspection and maintenance of structures, rigs, flowlines, risers, pipelines and corrosion protection systems.
- Cutting and disentanglement operations; recovery of items (misplaced equipment).

Also, as man is forced to search for energy resources in deeper water, subsea completed wells in association with buoyant crude processing structures will become more common for economic and logistical reasons. In addition to the activities listed above, manned intervention will be required for:
- Subsea wellhead servicing and repair.
- Wireline and through flowline in-tubing maintenance.
- Repair of manifolding and control systems.
- Valve operation and replacement.
- Inspection.
- Welding.

11

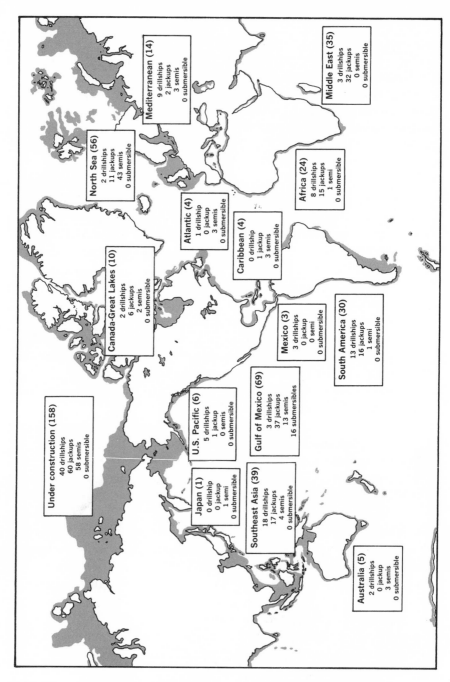

Mediterranean (14)
9 drillships
2 jackups
3 semis
0 submersible

Middle East (35)
3 drillships
32 jackups
0 semis
0 submersible

North Sea (56)
2 drillships
11 jackups
43 semis
0 submersible

Africa (24)
8 drillships
15 jackups
1 semi
0 submersible

Atlantic (4)
1 drillship
0 jackup
3 semis
0 submersible

Caribbean (4)
0 drillship
1 jackup
3 semis
0 submersible

Canada–Great Lakes (10)
2 drillships
6 jackups
2 semis
0 submersible

Mexico (3)
3 drillships
0 jackup
0 semi
0 submersible

South America (30)
13 drillships
16 jackups
1 semi
0 submersible

Under construction (158)
40 drillships
60 jackups
58 semis
0 submersible

U.S. Pacific (6)
5 drillships
1 jackup
0 semis
0 submersible

Gulf of Mexico (69)
3 drillships
37 jackups
13 semis
16 submersibles

Japan (1)
0 drillship
0 jackup
1 semi
0 submersible

Southeast Asia (39)
18 drillships
17 jackups
4 semis
0 submersible

Australia (5)
2 drillships
0 jackup
3 semis
0 submersible

Fig. 1-2.   Location of mobile drilling units world-wide, mid-1975.

The approach adopted for welding subsea is particularly significant in terms of the methods employed to overcome the problems imposed by the environment. In water depths up to say 150-200 ft, wet-welding techniques using divers have been used for some years. However, the joints so formed are not suitable for load bearing because of:

- Slag inclusions caused by the rapid quenching.
- The formation of martensite.
- The inclusion of hydrogen released in nascent form as a result of electrolysis of the water, leading to embrittlement.

Solving the problems. To overcome these problems, it is necessary to provide a dry environment, and the manner which this is done is dependent on operating depths and economics. In water depths suitable for diver welding, a local dry environment can be supplied by displacing water with inert gas from an enclosure surrounding the weld area.

The local dry environment so formed is at the same pressure as the surrounding sea from which the welder operates by inserting the welding torch into the enclosure. The enclosure is fabricated from transparent material to allow visual control of the operation. A hand-held movable unit is also available.

For greater depths where it is expensive and inconvenient to use divers for any lengthy operation, developments are in hand for a hyperbaric chamber with a dry, controlled environment. The welder in this case would be subjected to atmospheric pressure only, and can perform his task as if on dry land.

No one man can be the complete underwater engineer but a member of a multidisciplinary team combining a wide range of technologies and skills in this demanding environment. The need is here and will continue to grow.

2

Diving
Technology

MAN as a diver exposed to the physical conditions of the cold, wet pressures of the seas, has a leading role in the activity of underwater engineering. The task of placing a man safely at great depths is a combination of the physiological requirements and engineering techniques.

The physiological effects of diving are largely a result of first, the increase in pressure imposed by environment and second, the subsequent decrease in pressure during the period of releasing pressure ie. decompression sequence.

The increase in pressure necessitates the capability to equalize the pressures in the gas cavities of the body, the important ones being in the ears, head sinuses, and the lungs (and of course, also during the subsequent decompression). Normal diving procedures ensure that problems rarely arise from these sources.

Air is used as the breathing gas for diving to a maximum of 50m. The reason for this depth limit is largely due to an intoxicating effect caused by the nitrogen content of the air at these depths. The effect is often called "nitrogen narcosis" or "rapture of the depths." Australians call it the "uglies."

Whatever it is called, the effect is somewhat similar to alcoholic intoxication, and if air is breathed deeper than 50m, then the level of intoxication rapidly becomes dangerous and is therefore better avoided.

Special gas mixtures which may be used in deeper diving, such as helium and oxygen mixtures, can also introduce certain problems. For example, the human voice can be distorted to complete unintelligibility by helium, and special electronic devices referred to as "helium unscramblers" are required to return the speech to near-normal character.

While under pressure, the oxy-helium breathing mixture can cause a dangerous cooling of the body if it is not heated before being inhaled. This is due to the peculiarly high thermal conductivity of helium, coupled with its high density when under pressure. The sea itself is inevitably cold at depth, and the deep diver wears insulating clothing together with a means of providing supplementary heating around the body.

At very great depths, 300-600m, the increased density of the gases breathed can begin to limit the maximum breathing capability of a diver and thus his work performance. Extremely efficient breathing apparatus is therefore required at such depths in order to ensure the safety of the diver.

At the deeper end of this scale there may be adverse effects on the diver's nervous system, especially if the rate of compression was excessively fast. This effect has been called the "high pressure nervous syndrome," or HPNS, and consists mainly of a shaking of the limbs, making fine manual dexterity difficult.

Control of the percentage of oxygen in the breathing mixtures of deep divers is also essential. Although oxygen is the life-supporting gas, the quantity one breathes must be carefully controlled. Too much can be poisonous to the body, while too little leads to suffocation.

Controlling decompression. The decompression of a diver must be carefully controlled. The reason for this is that while under pressure, his body has absorbed a considerable extra amount of gas that entered via the lungs and blood system. During the decompression, this extra gas leaves the body again, returning to the lungs via the blood system and being vented out with normal breathing.

But if the decompression is too fast, the body may not be able to allow the gas to escape sufficiently quickly via the normal route, and the gas may appear as bubbles in the body. These bubbles can interfere with various functions of the body, in addition to perhaps causing aches and pains. These symptoms are often referred to as the "bends" or more correctly, decompression sickness.

15

Hot-water wet suit can make a diver more comfortable during shallow dives. Photograph, courtesy of Diving Unlimited, San Diego, California.

The number, size, and location of these bubbles determine the symptoms that may appear and these can vary from itching and pain in mild cases to paralysis and death in serious cases.

Thankfully, due to the adoption of proven decompression schedules by the commercial diving industry, even the mild cases are quite rare.

The treatment is relatively simple and extremely successful in practice. That is to recompress the diver in a chamber to reduce the size of the bubbles by stages, re-dissolving the gas, and then return very carefully to surface pressure.

Diving has advanced considerably in the last decade as a result of the offshore exploration for oil and gas but the underwater engineer must consider diving simply as the means by which a technician or engineer commutes to the work site (albeit a rather unusual site, possibly several hundreds of feet below the sea, and many miles offshore) and consider more importantly the task that the diver is to carry out.

Diving services are now available down to depths in excess of 300m and simulated dives (in research facilities) to 600m have been carried out. Open-water dives will probably advance to 600m in the near future, and this is considered by many to be the ultimate limit for "wet" diving. Several types of diving techniques are available and will be described in this account:

1. Surface orientated: Self-contained underwater breathing apparatus (SCUBA) and surface demand breathing equipment.
2. Bell diving: "Bounce" dives (short duration).
3. Bell diving: "Saturation" diving (long duration).
4. Diving from lock-out submersibles.
5. Habitat diving.
6. Atmospheric diving suits.

1. Surface Orientated—Self-contained Underwater Breathing Apparatus and Surface Demand Breathing Equipment.

These techniques are now well established but air as a breathing mixture is generally limited to depths of 50 meters (167 ft) due to the problems of nitrogen narcosis and its high percentage of oxygen.

Decompression can take place in the water for depths up to 20 meters (65 ft), but for greater depths some form of compression chamber, i.e. one man or two compartment chambers, would be required.

With surface orientated diving the divers can commonly be hindered by currents, wave conditions, and drag on the long hoses, with lifelines passing through the air/sea interface. The diver may be carrying equipment, and due allowance has to be made for the positive or negative buoyancy of this equipment.

Sea conditions rapidly become a limiting factor for handling a diver on a light staging or basket. As a result, the overall efficiency of this technique for long duration dives in areas like the North Sea is rather low and diving companies will probably use a diving bell even in the air ranges, when access for the bell system can be made available, sufficiently close to the job.

The free swimmer using SCUBA would be less handicapped by hoses but limited in duration by the amount of air that could be carried on his back. (North Sea offshore regulations require all SCUBA dives to be done with lifeline unless special exemption is given.)

2. Bell Diving (short duration—"bounce diving")

Usually diving bells are used for deep diving in the heliox range (mixtures of helium and oxygen), but are often used in the air range to overcome the problems of access for the diver through the air/sea interface and in carrying out decompression stops in the water.

In addition, the bell provides a shelter for the diver in which to rest, provides a store for tools and equipment, and can assist in communications. Below 50 meters the diving bell is used in conjunction with a lock-on deck compression chamber. The techniques then change and additional equipment is required for mixture heating and heating the diver's suit and life support systems for long exposure dives.

For anything other than brief inspection work or short jobs the technique of saturation diving is used. The figures below show typical decompression times for short duration or "bounce" dives:

Decompression Times for Bounce Dives

Depth, ft	Bottom time, minutes	Decompression time
246	30	2 hours
246	60	2 hours
305	30	3 hours, 20 min
305	60	9 hours
450	30	12 hours
450	60	12 hours

NOTES:
Preparation for a dive approximately 1 hour.
Bottom times limited to 60 minutes maximum or 120 minutes in emergency.
Normally rest between dives of 24 hours, in an emergency 12 hours.

As can be seen from the figures in the table, operating on "bounce" diving techniques, the bottom time at any depth is severely restricted by decompression. More bottom working time can be achieved using bounce-dive techniques if a second diving team is used while the first team is still undertaking its decompression. However, this necessitates the use of a second compression chamber for the decompression of the second team, independent of the first team. This technique is referred to as "back-to-back" diving. For depths beyond 100 meters and for long duration dives saturation diving techniques are employed.

3. Bell Diving (long duration)

Saturation diving. A diver works under increased ambient pressure, and his body tissues absorb the inert gas in the breathing mixture at a rate dependent upon the depth and the time spent at depth. Hence the decompression time (the time to release the absorbed gas safely) is a function of how much gas his body tissues have absorbed.

If the diver remains down long enough his body tissues are said to become eventually saturated with the inert gas; that is, no further gas is absorbed, and therefore no further increase in the

Biggest saturation-diving system built by COMEX, Marseille, France.

decompression time is required. It is possible to take advantage of this fact in commercial diving.

After the divers have completed their first dive, they return to the bell and instead of commencing decompression, they stay at the bottom pressure (or some intermediate pressure), and the bell is lifted back to the surface and locked on to the deck compression chamber which is also maintained at the bell pressure and filled with the breathing mixture.

20

Latest design diving bell for 1,000-ft depths. Photograph courtesy of COMEX.

The divers then transfer under pressure (TUP) into the deck chamber where they can rest, eat and live relatively comfortably until required to carry out another dive.

This way, using several divers, a service can be made available almost continuously with divers spending 2-3 hours or more on bottom. On completion of the task, the length of decompression time remains relatively unchanged as for a short duration dive. However, this technique does require a substantial amount of equipment in size and weight (see details).

The following list gives an outline of the typical equipment required to carry out saturation diving:

1. Diving Bell rated to suit the depth (say 200m) with internal heating, regeneration system, internal and external lighting and power supply.
2. Two, 2m diameter deck compression chambers complete with gas scrubbing unit.
3. Electrically operated bell handling winch, complete with

Diver returns to bell after completing job. Photograph courtesy of COMEX.

electro-mechanical cable (say 250 meters long) and multi-speed control.

4. Hydraulic "D" Frame, or travelling beam depending on access from the surface vessel for handling bell in and out of the water and to lock on to the deck compression chamber.

5. Control cabin fitted with helium unscrambler, communication system for the bell and divers, communication link to others, e.g. drilling superintendent, gas analyzers and gas mixture control system.

6. Gas compressor (electrically operated, explosion proof, membrane type) capable of handling pure oxygen.

7. HP air compressor/electrically operated, explosion proof.

8. Bottle banks for storage of mixed gases.

9. 400 bottles of helium and 50 bottles of oxygen.

10. Six complete sets of personal diving equipment, including heliox-demand breathing sets, masks with diver communications.

11. Six complete sets of SCUBA, wet suits, hand lamps, watches, compressor, and surface-demand equipment.
12. Diving stage for surface-orientated diving.
13. Constant-tension guide wire control system for positioning the bell and safety lift device.
14. One set of oxy-arc cutting and welding equipment, consisting of 210 meters of welding lead and oxygen hose and 100 lbs of cutting rods.
15. One set of hydraulic tools with underwater hydraulic power pack, impact wrench, brushes, etc.
16. Bell umbilical gas hose, length dependent on location of diving systems; i.e., height above sea level, depth of water, etc.
17. Electric or diesel powered winch for bell umbilical gas hose handling.
18. Two compression-chamber heating units.
19. Two gas-regeneration systems.
20. Six electrically heated undersuits, or six hot water suits plus heater unit and hoses and handling system.
21. Six electric gas heating packs.
22. One decompression sanitation module.
23. Diver held, low light, closed-circuit television system or bell mounted system, or remote controlled system.

4. "Lock-out" Submersibles

Instead of a diving bell the divers can be supported from a lock-out submersible (see Chapter 4, Underwater Vehicles).

5. Habitats

Sealab I. This was the U.S. Navy's first undersea living experiment and was staged 30 miles off Bermuda, between 20th July and 1st August, 1964. Four men lived for 11 days below 192 ft of water. Their habitat was a steel chamber measuring 40 ft in length by 10 ft in diameter. The habitat was self-contained except for electrical power, which was supplied by cables from the surface support ship.

The aquanauts (now an accepted term for the explorers of innerspace) could leave their steel home through two manholes located in the deck of the chamber, without intermediate locks, as the pressure inside the habitat was the same as the deep environment.

The principal accomplishment of this experiment was not the work performed, but the fact that men had lived in the open sea and under ambient pressure for an extended period of time (12 days). It highlighted areas of testing communications, and life support systems which required further research and development for man to penetrate safely into the sea.

Sealab II. This was to some extent an extension of the earlier work, occurring for 45 days from August to October, 1965. This experiment was held at a depth of 205 ft off La Jolla, California. Three teams of 10 aquanauts successfully lived at this depth in a specially built habitat measuring 57 ft long by 12 ft. It was in Sealab II that Astronaut N. Scott Carpenter turned Aquanaut, lived with two teams, remaining below for a total of 30 consecutive days.

Sealab III. The Navy's third experiment was lowered to 610 ft off San Clemente Island, a few miles off Southern California, on the 15th February, 1969. However, on the 17th February, in attempting to make the habitat ready for occupancy an Aquanaut died while swimming near the habitat. The experiment was immediately suspended and investigations carried out.

The experiment was to have placed on the sea floor five teams of nine men each for consecutive 12 day periods. Their mission was to carry out experimental construction and salvage work, oceanographic and marine biological research, and undergo a series of physiological and human performance tests.

Tektite. The General Electric Company designed, engineered, fabricated and tested an undersea dwelling, thereafter called a habitat. This project was named Project Tektite (a name it is reported that everyone thought horrible) and on the 15th February, 1969 four scientists/divers (three from the Bureau of Commercial Fisheries and one from the U.S. Geological Survey) entered the habitat in 47 ft of water to begin a 60-day stay.

Tektites contributions to the habitat concept were:

1. The length of time that the subjects remained under pressure.
2. The use of nitrogen rather than helium as the inert dilutent of the breathing mixture, so that any loss of performance of the divers due to nitrogen saturation could be recorded.
3. Emphasis on scientific (rather than technological) tasks.

The Underwater Laboratory (UWL). The Underwater Laboratory (UWL) Helgoland was designed for marine-biological research and for medical-psychological tests on men working under water and was positioned on 28 July, 1969 in the North Sea.

The UWL Helgoland was constructed with a pressure-proof body of cylindrical shape, the length being approximately 9m (29 ft 6 in) with a diameter of approximately 2.5m (8 ft 2½ in). See Fig. 2-1. The hull stands on four adjustable legs which are fitted with two ballast weights for stability. The pressurization capacity of the underwater complex was designed for an equivalent water depth of 100m (328 ft). The total volume amounts to approximately 43m³ (1,520 ft³). The weight of the UWL including ballast amounts to approximately 60 tons.

The complex consists of two separate compartments, one wet-room and one work-room. The frontal walls of the UWL are fitted with two beds each. The wet-room is fitted with an oval hatch through which the Aquanauts enter the water. This hatch has a length of approximately 2.4m (7 ft 10 in) and was designed to compensate for changes in water level as caused by tides.

On each side of the underwater complex are ballast tanks which may be flooded and released, thus facilitating swimming, transportation, and a comparatively simple change of stationing. This ballast also provides the necessary negative buoyancy for stability under water. The straddled legs have been fitted with boxes to take up additional ballast for constant level positioning on the seabed.

The power source of the UWL and the breathing gas supply are provided by a supply buoy on the water surface, which is moored adjacent to the Underwater Laboratory. The buoy enables constant communication with the outside by radio and television. An additional cable provides for teletype and telephone communi-

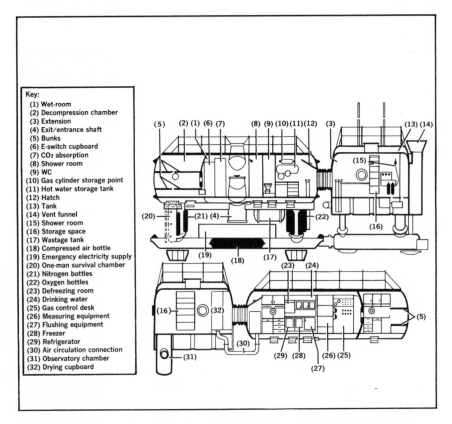

Key:
(1) Wet-room
(2) Decompression chamber
(3) Extension
(4) Exit/entrance shaft
(5) Bunks
(6) E-switch cupboard
(7) CO_2 absorption
(8) Shower room
(9) WC
(10) Gas cylinder storage point
(11) Hot water storage tank
(12) Hatch
(13) Tank
(14) Vent funnel
(15) Shower room
(16) Storage space
(17) Wastage tank
(18) Compressed air bottle
(19) Emergency electricity supply
(20) One-man survival chamber
(21) Nitrogen bottles
(22) Oxygen bottles
(23) Defreezing room
(24) Drinking water
(25) Gas control desk
(26) Measuring equipment
(27) Flushing equipment
(28) Freezer
(29) Refrigerator
(30) Air circulation connection
(31) Observatory chamber
(32) Drying cupboard

Fig. 2-1. General arrangement drawing of the Underwater Laboratory, Helgoland.

cation with the shore control station. The Underwater Laboratory is provided with an independent means of life support and power for an emergency period of two weeks.

Stock replenishment is performed by way of a pressure container. Supplies may be stored in an underwater depot immediately adjacent to the laboratory in order to economize on space in the UWL.

The atmosphere within the UWL is continuously controlled by metering units. Exhaled carbon dioxide is absorbed by canisters, the exhausted oxygen is continuously replenished, and the temperature as well as the humidity are controlled automatically. The

26

atmospheric conditions are transmitted to the shore control station by way of television.

A special feature of this project is that the decompression schedules are performed within the Underwater Laboratory, and that automatic supply of power and breathing gas is provided from an unmanned buoy.

Numerous safety and rescue facilities have been provided: such as a "one man" pressure chamber which can be released from the UWL and subsequently recovered on the surface by a helicopter or boat, which are constantly stationed near Helgoland.

In cooperation with other prominent firms Messrs Drägerwerk have solved the different problems which are encountered in the technical equipment of the various fittings, such as a special WC, a hot water shower and a complete electric kitchen (galley) including an emergency breathing system. Prior to final delivery each single item was checked by simulated diving tests.

Together with the UWL Drägerwerk supplies a water "Igloo" which is intended as a "satellite" to the big brother UWL.

The first international, long-term cold water saturation scientific diving program was conducted from September through November 1975 in 110 ft of water, 10 miles off Rockport, Mass. using the West German Underwater Laboratory "Helgoland." The program was coordinated by the National Oceanic and Atmospheric Administration (NOAA) and over sixty participants were involved from the United States, West Germany, and Poland. The work involved three 4 man teams of scientists who completed 164 man-days of saturation bottom time. They completed an ecological study of a herring spawning site, studied fish behavior and trapping methods, and initial tests of a hydroacoustic calibration system.

6. Atmospheric Diving Suits

To overcome the physiological problems of diving the concept of the atmospheric diving suit has been evolved. The famous suits in the 1930's by Neufeldt and Kuhnke, however, became difficult to operate below a few hundred feet and were reduced to little more than observation chambers. Even the suits of the Italian, Roberto

27

Galeazzi, which are still available, tend to become ineffective before their operating limit of 600 ft.

On 26th October 1935, British diver, Jim Jarratt reached the wreck of the Lusitania lying in over 300 ft of water off the Old Head of Kinsal, County Cork. The dive was carried out in a diving apparatus invented by Joseph Salim Peress. Unfortunately, rough seas prevented any salvage operation on the ship. The Peress suit, probably the most advanced armoured diving suit of its time, faded into obscurity.

Recently a British-owned company D. H. B., with support from the National Research Development Corporation, has returned to the view that deep diving should be an engineering problem not a physiological problem.

The suit has been nicknamed "JIM 2," after its diver, Jim Jarratt (JIM 1). The suit will carry the operator at atmospheric pressure to depths of 1300 ft. D.H.B. have established an operating agreement for the use of these suits with Oceaneering International.

The ADS—"Jim 2"

The ADS is a self-contained unit carrying a life-support system consisting of two independent units with a total endurance of up to twenty hours. Two separate supplies of oxygen are carried externally within a back pack. After adjustment at the beginning of a dive, compensation for the varying oxygen requirement of the operator occurs automatically. Monitoring of the atmosphere within the suit shows that oxygen and carbon dioxide partial pressures remain within tolerable limits. After long dives the atmosphere remains wholesome, with little trouble from condensation.

The suit will accommodate a wide range of operators of differing physiques; normal clothing may be worn under an overall. Adequate room inside allows operation of all internal controls, and stowage containers may be carried internally for additional equipment; for example, notebooks, camera and, if required, food and drink.

Communication with the operator is maintained by telephone; to permit autonomous untethered operation, a through water communication system could be fitted.

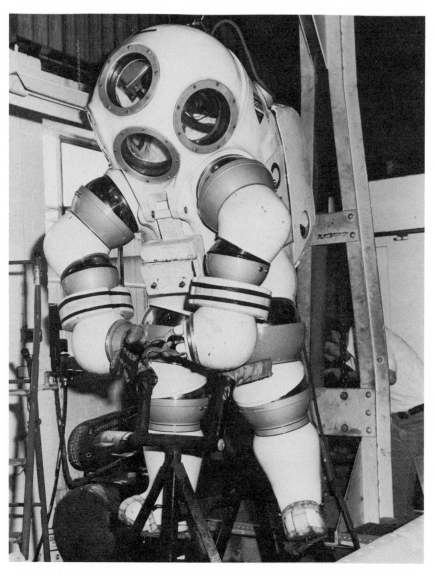

One-atmosphere diving suit for possible use by British Petroleum Corp. in subsea operations.

Manipulators and tool adaptors have been developed which can be interchanged rapidly by means of a simple bayonet fixing. Dexterity trials have included work on standard fittings such as shackles, hand wheels and nuts and bolts.

The operator can pick up objects from the bottom, climb a ladder, and negotiate moderately severe gradients. The suit is stable in the upright position, but the operator can at will lie prone or supine and recover to a normal walking attitude easily. Visibility through the four large plexiglass windows is excellent.

Lead ballast of approximately 150 pounds weight is required to trim the unit for work; the negative buoyancy of the suit is adjusted according to bottom conditions with the range 15-50 pounds weight. The rear ballast is the primary drop ballast; when released the suit ascends at approximately 100 feet per minute. After ballast has been jettisoned the suit floats at the surface in an upright stable state.

The unit is transportable by road and air. In use it requires approximately twenty square feet of deck space for stowage and access, and a hoist with a capacity of 1100 pounds weight. The suit is lowered into the water by a main lifting wire which is then detached and lowering is completed using the load-bearing telephone cable.

It is considered that a four man team can run an eight-hour-day work cycle with reserve for working through. Support equipment consists of a standing frame for the assembled suit, and packs containing communications apparatus, supplies of absorbent for carbon dioxide, and spares and tools. Routine maintenance can be undertaken by the composite diver/mechanic team.

Diving Services

Diving services are provided for a wide area of activity and involve the various techniques and equipment mentioned previously. The following provides an idea of the extent of diving services:

1. Military diving; e.g., torpedo recovery.
2. Inland waterways.

3. Docks and harbors.
4. Offshore exploratory rigs, installation and construction of platforms and pipelines, inspection and maintenance, and welding habitats.
5. Salvage and recovery.
6. Fisheries research.
7. In-water maintenance of tankers.
8. Research habitats.
9. Lock-out submersibles.

Typical offshore diving tasks.   These include:

Inspection/Survey
Visual Inspection, Photographic Inspection, TV Inspection, Bottom sampling, Bottom probing, Tape Survey.

Preparations
Removal of debris, rock breaking.

Offshore Platforms
Inspection visual and TV.

Removal of marine growths.

Checking of scour, installing and maintaining scour prevention system.

Cathodic protection inspection, repair and renewal.

Non-destructive testing.

Drilling Rigs
General assistance with subsea drilling operations.

Emergency repair service.

Pipelay Barges
Controlling stinger profile.

Checking pipe profile.

Pipe burying.

Construction Work
Control of platform installation works.

Cutting of piles, and buoyancy systems.

Grouting.

Anchor.

Single buoy mooring installations.

Pipe connections and repairs (hydrocouples/flanges).

Underwater welding, wet and habitats.

The "man in the sea" as an aquanaut is now and will continue to provide the front-end support for underwater engineering.

His services are cost effective to 300 meters and soon will be available to 600 meters.

However in the range of 300-600 meters his role will change to that of making the difficult first surface to seabed contact and as a backup on other techniques.

Beyond 600 meters the engineering ability of man is required to perform tasks remotely or in atmospheric conditions. Many feel that the 600 meters depth is the "working" limit for the "wet" diver.

3

Inspection and
Maintenance

INCREASINGLY large structures are being placed in the sea, and sub-sea pipelines of ever-greater diameters and lengths are being laid. Although this is a world-wide trend in the oil and gas industry, the lead is being taken by the work of North Sea constructors to meet the requirements of deeper water, greater climatic extremes and the practical problems of installation. All structures will require a degree of routine inspection and maintenance as well as rectification of the defects that will inevitably occur.

It is perhaps the scale of the task that is most daunting, a modern platform such as Graythorpe I for B. P.'s Forties field contains 30,000 tons of steel and stands in 400 ft of water. Platforms for 500 ft depth are under construction, and structures greater depths 1,000 ft are planned. Clearly the inspection and maintenance requirements will be immense. See Fig. 3-1.

Similarly, gravity structures of concrete such as lighthouses and offshore storage tanks may need to be inspected for scour around the footings. Pipelines will require inspection along their length, for corrosion, leaks, and integrity of weight coating. Where a pipeline is intended to be buried, the depth of cover will need confirmation. This will require accurate means of position-fixing. Since, in general, only pipelines below fishing depth will be left unburied, this will constitute a major inspection effort.

The largest North Sea pipeline is 34-in. diameter, currently (1976) being laid, but larger pipelines of 36-in. diameter are

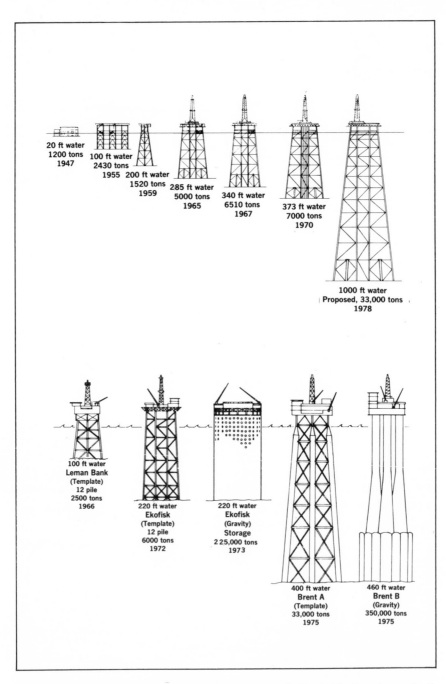

Fig. 3-1. 27 years of platform development, Gulf of Mexico (upper). 8 years of platform development, North Sea (lower).

planned. These will each be about 100 miles in length and will be laid in water depths of the order of 500 feet.

It may be said that due to the scale of the task and the difficulties of access, inspection would take three times as long underwater and cost five times as much as compared with a similar land based operation. Inspection would cover both non-structural aspects and range from a purely visual inspection to the use of non-destructive testing equipment.

To carry out these inspections, remote controlled TV, divers, and underwater vehicles could be used.

Non-Structural

Visual. The purpose of the visual inspection would be to assess the general state of a structure, particularly with regard to marine growth and protective coating. This type of inspection would be carried out before commencing any other form of inspection, in order to plan the program of more detailed inspection

Diver using a hand-held still camera on a North Sea structure.

and/or maintenance. For a structure, this would involve observation by a variety of techniques.

First, a diving bell operating at atmospheric pressure can be lowered from the surface around the periphery of a structure, allowing the observers to make a general inspection. Alternatively, if close TV surveillance is required, divers can pressurize the bell, enter the water, and examine the structure closely, with the aid of hand-held cameras. The length of a diver's umbilical would allow him to move a little way inside the structure and examine areas that a bell or submersible would not be able to cover.

Secondly, remote controlled submersibles equipped with lights and video TV cameras may be used. There are several types of vehicle available, some of which are very small and compact (Fig. 3-2). This gives them great maneuverability, but detailed photography is hindered by their natural instability in mid-water regions. The unit is designed to be as small as possible in order to minimize the drag forces subjected to it by the sea currents.

Diver using a hand-held still camera on an underwater pipeline.

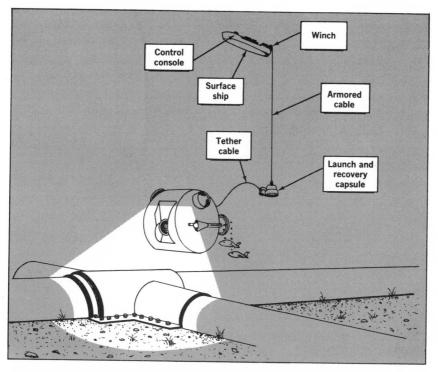

Fig. 3-2. Remote controlled pipeline-inspection vehicle.

Scour and debris. Inspection for scour will mainly be visual and related to the foot of a structure. The main problem, therefore, is access due to depth, and is an activity where a submersible operating at atmospheric pressure may well be more effective. Similarly for debris, the general inspection could be carried out by a submersible fitted with side scan sonar and closed-circuit television.

Remote inspections using Closed Circuit TV (CCTV) have been undertaken of the Ekofisk storage tank. Although demonstrating the viability of the technique, these showed the need for incorporating reference marks into structures, enabling the operator to locate specific points and estimate the attitude of the camera. Analysis of recordings made to monitor structure life will be more relevant if it is certain that successive examinations are of particular, accurately located features.

Similarly, on pipeline surveys using submersibles, a record should be maintained by still photography or video tape. At present for detailed records still photography is recommended. Stereophotography has also been used.

Removal of marine growth. This can be a very time consuming chore for divers, but is very necessary to ensure that the drag forces on a structure do not exceed the design limits. An allowance in the design is needed for this growth, which increases the overall diameter of a member and, thereby, the drag force. For example, allow 6 in. thickness on an 8-ft-diameter tube.

A wide range of pneumatic and hydraulic hand-held tools, effective down to 50 meters, are available for the diver such as: (1) powered scrapers, (2) needle guns, (3) nylon and steel rotary brushes, (4) chisels and grinding wheels, and (5) water jetting.

These could be operated from an underwater power pack or, for large vertical areas such as the sides of tankers, remote controlled units have been employed. Marine crustations are damaging to structures and are particularly difficult to remove, for which needle guns or water jets are required.

A facility exists in the Canary Islands for in-water cleaning of Very Large Crude Carriers (VLCC's). In the case of fast ships, particular attention is paid to removing goose barnacles (conchoderima aurita) from the flat of the bottom. Growths of this species have been recorded totaling 60 tons on the underside of a VLCC causing a speed reduction of as much as 1-1.5 knots. Underwater cleaning vehicles such as BRUSH BOAT, Fig. 3-3, and developments of remote controlled inspection vehicles such as SCAN, Fig. 3-4, incorporating a bottom scarping capability, will be useful here, and these may also have application to other kinds of structure.

Corrosion and protective coatings. During the course of a general survey, a check would be made on the cathodic protection system for anode wastage or loss, and the distribution of the protective potential.

The loss of metal from the structure due to corrosion would

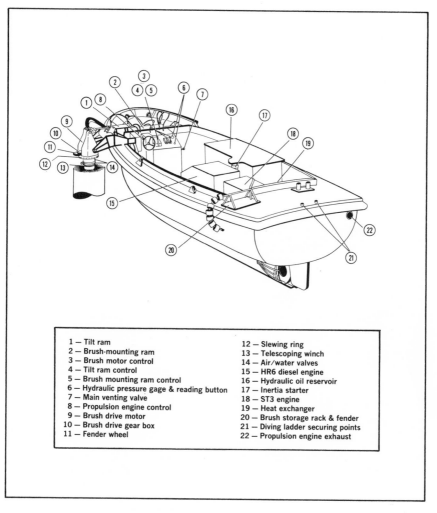

1 — Tilt ram	12 — Slewing ring
2 — Brush-mounting ram	13 — Telescoping winch
3 — Brush motor control	14 — Air/water valves
4 — Tilt ram control	15 — HR6 diesel engine
5 — Brush mounting ram control	16 — Hydraulic oil reservoir
6 — Hydraulic pressure gage & reading button	17 — Inertia starter
7 — Main venting valve	18 — ST3 engine
8 — Propulsion engine control	19 — Heat exchanger
9 — Brush drive motor	20 — Brush storage rack & fender
10 — Brush drive gear box	21 — Diving ladder securing points
11 — Fender wheel	22 — Propulsion engine exhaust

Fig. 3-3. The Brushboat underwater cleaning system.

also be determined using underwater ultrasonic equipment to measure metal thickness at points where serious corrosion is suspected or indicated.

Protective coatings have considerable limitations; no anti-fouling paint providing protection for more than two years, and 15 months is more likely. Wax coating techniques are at an experi-

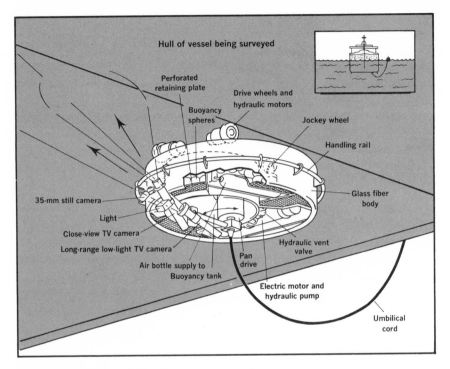

Fig. 3-4. Remote controlled hull-inspection system.

mental stage, but this in any event is unlikely to provide benefit for longer than four years.

In the shipping world, the buildup of marine growth reduces ship speed and increases fuel consumption by an estimated cost of $120 million per year for British shipping alone. Considerable research effort is currently being applied to finding ways of preventing marine growth forming, or of removing it, once formed. These also have application to offshore structures. There is no guaranteed method of preventing growth, although it can be restricted in the short term by the anti-fouling coatings.

Small isolated areas of damaged paint can be re-established by hand using underwater paint, but this is a slow job (rate 15 sq ft per hour). As the life of the anti-fouling protection is small in relation to the life of a platform, there is no virtue in providing this form of protection.

Care must be taken in using powered underwater brushing to ensure that it removes marine growth but does not cause damage to the existing paint film. The technique can be ineffective against hard shell, such as barnacles.

Structural

The following techniques are available for carrying out non-destructive testing:

Ultrasonics. Compression-wave mode of propagation is used for wall thickness surveys of exposed submarine riser bends and for internal scour and pitting. Ultrasonic techniques as yet cannot be guaranteed below 50 meters depth, and further development to extend the depth range will be required.

A satisfactory probe has been the Wells Krautkramer probe, due to several features of its design. The most important of these is the use of barium titanate as a standard crystal. The two crystals are totally encapsulated and bonded to a perspex face forming the scanning surface at one end. The power and detector cables enter the other end of the 6″ long probe.

Magnetic particle. By the use of proven magnetizing currents, a circular magnetic field, correct ink mixes, and a fixed standard or U/V lamp, satisfactory crack detection can be made. Several methods are in use underwater and these vary, mainly in the method of achieving magnetization. The method most widely used is the use of a probe. To enable this to be operated efficiently with maximum voltage drop, the probe itself is made of 1-in. diameter Muntz metal.

This was selected to ensure a good, hard, contact surface not liable to wear and with maximum electrical resistance. The 1-in. diameter probes terminate in flat chased edges which take the contour of weld circumferences and ensure maximum electrical contact. Both probes are set in an epoxy handle some 5 in. apart. The handle incorporates a magnetic-reed operating switch which controls a contactor in the transformer supply unit. Thus the trained operator can avoid applying the energized probe to the surface under inspection and causing localized arcing.

The whole transformer unit is mounted in a stainless-steel underwater housing capable of operating down to 450 ft, which can be supported on an adjacent cross brace close to the node under examination. The surface control unit contains indicator lamps in series with both the ultra violet lamps and magnetic probes, to provide evidence that the system is functioning.

An ammeter is fitted in the primary coil circuit, and is mounted on the control panel. This enables the operator's surface staff to ensure that the magnetizing current is of the minimum required to produce the 30 oersteds needed for magnetization. A secondary proof can be obtained by using a clip-on ammeter on the probe cables prior to the equipment going underwater.

Radiography. There has as yet been very little call for this work underwater, and further development is required of either X-ray or Gamma-ray techniques. An advantage of radiography is that it is not pressure sensitive; however it is unlikely to be used on an irregular shape, such as a node, due to the difficulty of placing films. Its particular application would appear to be the examination of submarine pipelines and marine risers which are coated with concrete and bitumen.

Eddy current. This method, which shows promise, is now being developed for underwater use. Its major land-based use has been for the examination of hull and system welds in nuclear submarines.

Maintenance

The main activities of maintenance are cutting and welding on structures, and the prevention of scour on pipelines by placing concrete and synthetic mattresses. Several methods are available for cutting underwater, each with its own particular limitations, and some under continuing development:

1. *Gas cutting.* Torch cutting of iron and steel on the surface or underwater is accomplished by burning or oxidizing the material. The material to be cut is raised to red heat by one of several methods, and oxygen to enable burning is directed at that point.

Underwater gas-torch cutting uses acetylene or hydrogen for the fuel gas, although acetylene is very unstable at pressures in excess of 15 psi, and so is only used in depths of less than 25 ft. It is possible to use hydrogen at greater depths, but the hydrogen flare is not as hot as acetylene.

Although gas torch cutting underwater is possible and practical, it is seldom employed in offshore work due to the bulk of gases required and the high degree of technical skill demanded of the diver.

2. *Oxy arc cutting.* This is the most widely used method to date for cutting underwater. This process produces tremendous heat (about 6500° C) with an electric arc to preheat the metal. As with gas cutting, when the steel is heated, a high-pressure jet of oxygen is impinged on the heated steel to complete the cutting process. However, oxy-arc cutting leaves a very rough and ragged edge.

Equipment required is an arc welding electrical supply, a source of high-pressure oxygen, an oxygen pressure regulator, welding and earth loader, an earthing clamp, oxygen hose, a single-pole knife switch of sufficient size and capacity to carry the required current, a supply of tubular steel electrodes, and an oxy-arc cutting torch.

Thicker steel requires higher current (normally 300 amps) as do the longer leads necessary for work in deeper waters. It is possible to cut with either ac or dc current, but dc is preferable because of its lower potential for shock and reduced corrosive action on the torch and metal parts of the diving gear. When using ac currents, any metallic part of the diving gear that will come into contact with the diver must be insulated. The welding machine frame should be earthed to the barge or surface support unit. Oxy-arc cutting is available to 150 meters depth.

3. *Thermic lance.* A typical lance is a ⅜-in. diameter pipe, 10 ft, 6 in. long packed with a number of rods of different metal alloys, such as aluminum, magnesium, thermite and steel. High pressure oxygen is forced through the pipe and the lance, once ignited, burns with tremendous heat e.g. 10,000° C. A 10 ft, 6 in. bar lasts about 6 minutes, and it is this feature that poses the

greatest problem underwater. A number of bars can be coupled together for longer burning, but the extra length makes handling extremely awkward.

The advantage of the burning bar is that it will burn or melt through almost anything—steel, nonferrous metals, rock, and concrete. Its greatest use is for very heavy steel components such as shafting, and it could also be used for cutting pipelines quickly without the necessity of removing the concrete cover and protective coating.

The thermic lance must be ignited by some outside source of heat, such as a burning torch or, underwater, by oxy-arc torch. Because of its awkwardness and the high rate of oxygen consumption, the burning bar is practical only for very heavy steel or other materials that cannot be cut with the oxy-arc torch. At present the method is being developed to be acceptable down to 100 meters.

4. *Thermal-arc.* A decided improvement over the thermic lance is the thermal-arc cutting equipment. The principle is the same but instead of a rigid bar, the consumable agent is a tough, flexible plastic-covered cable which comes in 100-ft lengths. The cable, as opposed to six minutes for the bar, has an average burning time of approximately one hour. Because of its flexibility, it is less awkward and, therefore much easier to control and use. The cable is hooked up to an oxygen hose and welding cables similar to oxy-arc equipment, but the electrical circuit is used only for igniting the thermal cable and is not used during the actual burning.

The most widely used equipment is manufactured by Clucas and is under development for operation to a depth of 150 meters.

5. *Arc plasma.* This is a new technique which may well have an application underwater for cutting and welding. Plasma is the term used to describe a body of gas in which atoms are ionized by raising it to a high temperature. Unlike gas in the normal state, the ionized form is an electrical conductor and may be influenced by magnetic and electric fields.

It is understood that temperatures in excess of 30,000° C can be achieved, that is, more than three times the temperature of a conventional welding arc. A simple arc-plasma torch consists of

44

a tungsten-rod cathode and a water-cooled copper nozzle anode positioned in an insulated body, through which gas is fed. This is set up by the diver, but is remotely controlled, indicating a need to develop the equipment into a more directly controllable form.

Welding techniques. Welding underwater is still very much in the development stage and the processes of welding under pressure and in the "wet" have yet to be fully understood. High standards for underwater welding have yet to be achieved. Although "wet" welding has been carried out in depths of 50 meters, this has yet to be widely accepted in the field.

Early attempts at underwater welding used conventional arc equipment with a special rod coating, which by generating a gas bubble around the arc, would reduce the quenching effect of the surrounding water. More recent attempts were with the metal-

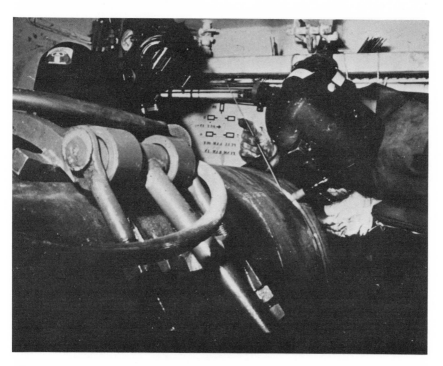

Fig. 3-5. Lines joined in dry underwater habit. Photograph courtesy of Ocean Systems.

inert-gas (MIG) process. This also depends on maintaining a bubble—this time, of inert gas. All such attempts have until recently produced a marginal weld rarely approaching acceptable standards. At greater depths the bubble, being compressed to smaller size, had even less effect and in unfavorable welding positions became difficult to control.

Recent developments have attempted to provide a dry environment around the weld in a welding habitat (see Figs. 3-5 and 3-6), and with this technique successful pipeline welds have been made in 540-ft-depth offshore Louisiana. The cost and difficulty of this process, with hyperbaric diver-welders and welding habitat, restricts its use however.

Extending the dry-environment principle to smaller and cheaper equipment suitable for one-man operation was a logical next step.

Fig. 3-6. Dry habitat for underwater pipeline work. Photograph courtesy of Ocean Systems.

This led to the Hydrobox and Portable Dry Spot (PDS), both offered by Sub-Ocean Services. In the Hydrobox, an inverted transparent bell is fitted over the weld region, dewatered by filling with inert gas from below, and welding undertaken by a diver inserting a hand-held MIG-type gun into the dry environment so formed. (See Fig. 3-7.)

The diver can watch and control the process through the transparent sides of the bell and has an eye-protecting dark visor fitted to his helmet.

Fig. 3-7. Application of Hydrobox process. Photograph courtesy of Sub Ocean Services Inc.

47

The PDS uses the same principle but on an even smaller scale. The welding gun, also of the MIG type, has a small dry environment built around it and is kept dewatered by a continuous inert gas supply.

The diver observes progress through the rear window and the whole unit is moved continuously as the weld progresses. Both processes can produce welds reliably to API 1104 Code Standards.

Explosives

Historically, explosives have been extensively used underwater, mainly for demolition work of various kinds, and in relatively unsophisticated ways. Developments of explosive uses for military and aerospace purposes have more recently been used to extend the range of operations possible by using small charges in more efficient ways. Thus, blasting of specific targets is now possible without causing damage to other quite-nearby objects.

Explosives may be divided into three classes. The first, low explosives of the gunpowder or cordite type, explode relatively slowly and exert a pushing effect rather than a shattering blow. These are, however, seldom used offshore.

High explosives detonate almost instantaneously, creating powerful shock and pressure waves, and are used for most military and commercial purposes. To reduce costs, commercial explosives often consist of various mixtures of a high explosive and a blasting agent, the third explosive class, such as ammonium nitrate.

A typical commercial explosive consists of nitroglycerine absorbed in an inert filler such as sawdust or kieselguhr, which also reduces the nitroglycerine's great sensitivity to shock. In this form it is known as dynamite, which is used extensively offshore. However, it is not waterproof and deteriorates underwater, depending for its useful life of 24 hours submerged, on its wrapping.

For complex charge-setting operations which extend over a period of days or where there is danger of delay (for example by bad weather), gelatine dynamites are used instead. These are dynamite mixtures plasticized and rendered inherently waterproof by the addition of various agents. Their higher cost is justified by their greater flexibility where delays are possible.

48

Blasting agents are the cheapest of explosives and are most suited for very large charges such as those needed for wreck dispersal. As they have poor water resistance, they depend on waterproof packaging to maintain underwater effectiveness. This class of explosives cannot be easily detonated, and a stick of dynamite in each charge is needed as a primer.

Detonators or blasting caps are required to initiate the explosion of charges. They consist of a small aluminum or copper tube closed at one end and containing a sensitive charge such as mercuric fulminate. Or, more usual in British-made detonators there is a mixture of lead azide, lead stiffnate, and aluminum powder, and known as "ASA" from "azide-stiffnate-aluminum."

As shown in Fig. 3-8, an electrical bridge surrounded by a heat-sensitive mixture is sealed into the open end of the detonator tube, the wires leading to the bridge being twisted together or fitted with a crimped-on shunt to prevent accidental explosion by stray electrical charges, static, or radio/radar emissions.

In modern practice, it is usual to connect the detonator to the explosive charges by a length of explosive cord known as "Primacord" or "Cordtex." This has several advantages. The diver does not have to handle the detonator, which can be attached to a length of the cord brought to the surface. Several charges can be detonated simultaneously, even though widely separated, due to the fast (21,000 ft per second) rate of detonation of the cord. Primacord can also be used as an explosive in its own right, a few turns exploded around a pile clearing it of marine growth.

The sequence of operations in carrying out an explosive demolition begins with placing the charges. These may need priming

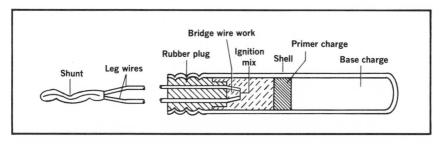

Fig. 3-8. Electric blasting cap.

with a stick of explosive as described, and could have a length of primacord attached when being made up at the surface. With the charges in place the various lengths of primacord are then joined together and a leader run to the surface. A detonator is attached to the end of the leader and the exploder wires to the detonator. These are then dropped overboard and paid out as the vessel moves away from the blast area.

Before the charge is fired, a check for electrical conductivity is made with a special low-current galvanometer and, if approved, the exploder, or blasting machine, is connected to the exploder wires and operated.

Shaped charges. The latest developments in underwater explosive technology have made use of military and aerospace industry experience with shaped charges for anti-tank warheads and missile stage-separation. (See Fig. 3-9.) These can be made up locally using, for example, a coffee tin with a conical metal former, filled with repacked explosive as a hole-puncher, or sheet explosive of the Du Pont Detasheet type, bonded to a metal tube as a linear cutting charge.

However in their most sophisticated form they are accurately pre-made for a variety of purposes. These include circular, for cutting piles, linear for various purposes and an array of three linear charges held by a former to fit over and cut I-beams. Additional forms can be made to order, and, although very expensive, their great advantage is in diving time saved.

The simplest form of circular cutter is for outside attachment (Fig. 3-10) and consists of a tube containing explosive packed behind a shaped former and bent to fit precisely around a pipe of given size. They are made in two exact semi-circular sections hinged together and with a latch for connection around a pipe. All such charges are detonated with primacord attached at a special fitting, and the resulting cut resembles that of a gas torch.

An inside circular cutter, which may be lowered inside a pile, consists of a shaped explosive charge contained in an inflatable tube and carried on a former. This is lowered to the desired depth, air pressure is applied to inflate the tube and hold the explosive at the correct stand-off distance from the surface, and the charge exploded. See Figs. 3-11 and 3-12. Linear or straight charges are also packed in tubes with plastic feet to aid correct alignment and stand-off (Fig. 3-12), and can be assembled in a variety of ways such as for I-beam cutting (Fig. 3-13).

Small versions are also available for chain or cable-cutting, Fig. 3-14.

Explosive stakes have recently become available for such purposes as fixing pipelines to the sea bed. This requires the pipeline to be laid with a fixed tangential flange flat on the sea bed. An explosive stake is driven through a hole in this, using a setting tool operated from above surface, to pin the pipeline in position. Later, a small charge built into the bottom of the stake is exploded, opening the lower half of the stake into three petals and securely anchoring the stake and pipeline to the sea bed.

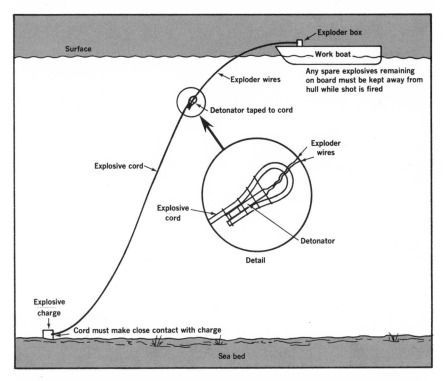

Fig. 3-9. General method of firing underwater charges in safety.

51

Preparation of a shaped charge for removal of protective concrete coating.

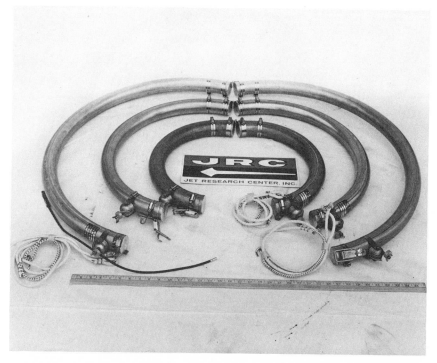

Fig. 3-10. Outside circular cutters. Photograph courtesy Jet Research Center, Inc.

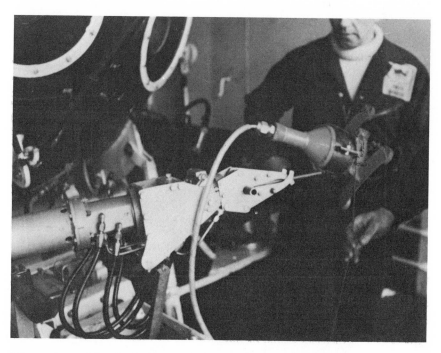

Charge for cutting holes in pipelines held by a manipulator.

Fig. 3-11.   Inside circular cutters, Jet Research Center, Inc.

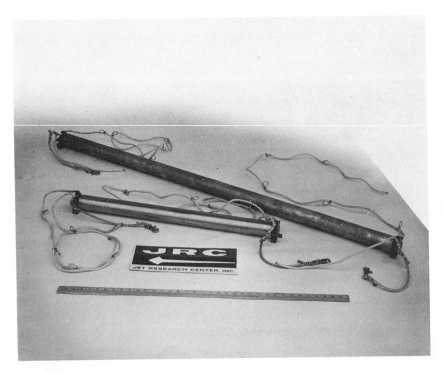

Fig. 3-12.   Straight cutters, Jet Research Center, Inc.

54

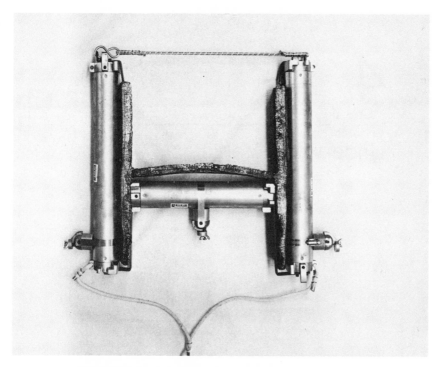

Fig. 3-13. Structural beam cutter, Jet Research Center, Inc.

Fig. 3-14. Cable and anchor chain cutters, Jet Research Center, Inc.

4

Underwater Vehicles

A FLOATING vessel can be submerged in two different ways.

1. *Statically,* by reducing the displacement or increasing ballast.

2. *Dynamically,* by applying a downward force.

To reduce the displacement the vessel has to be equipped with ballast tanks which can be flooded. Dynamic submersion can be achieved by operating horizontal rudders (only operative if the vessel is in motion) or by vertical propellers. In practice submersion is made by a combination of these techniques.

The most difficult problem that had to be overcome in the development of submarines was the problem of depth-keeping. Lack of controlled buoyancy caused the vessel to sink, hitting the bottom or collapsing if in really deep water. Without sufficient engine power it was impossible to develop the required downward thrust on the horizontal rudders or vertical thrusters. In addition early submarines were unstable longitudinally owing to their low metacentric height.

The hand-propelled submarine was never more than a death trap, and the major breakthrough came with the advent of electrical motors and the internal combustion engine. Conventional submarine power reached a peak when the British Submarine H.M.S. Andrew made the submerged trip from Bermuda to England in 1953 of 2,500 miles. The first atomic powered submarine, the U.S.S. Nautilus (SSN571) was launched in 1955.

Nautilus broke every existing submarine record for speed, endurance, and range. Her first voyage from New London to San

Juan, Puerto Rico was more than 1,381 miles completed in 84 hours submerged. In many ways Nautilus can be considered the first real submarine, as previous underwater vessels had really been surface vessels capable of submerging for only short periods. On active patrol it was common for them to move as slowly as 1 knot when submerged to conserve battery power or even to drift with all engines off.

A further achievement that certainly placed undersea travel in a new era was the voyage of Triton, a twin-reactor-class submarine which traveled submerged and undetected around the world. This submerged journey, following the route of Ferdinand Magellan (a Portuguese explorer in 1615) covered 41,500 miles in 84 days at an average of 500 miles a day.

The distribution of nuclear and conventional submarines worldwide is as shown in Table 4-1.

From these weapons of war have evolved the designs and manufacture of pressure hulls, heavy duty batteries, escape systems, and buoyancy control systems.

Submarines and submersibles are now used for scientific research missions, for sampling and surveying the seabed on the routes of pipelines, and laying and burying submarine telephone cables, torpedo, and general recovery.

With the launching of large submersibles such as the R/V Ben

TABLE 4-1
Submarine Distribution

	UK	USA	France	USSR	Japan	China	East Germany	Italy
Missile Submarines								
Nuclear	4	41	2	87	—	—	—	—
Conventional	—	—	—	50	—	1	—	—
Attack Submarines								
Nuclear	7	58	—	26	—	—	—	—
Conventional	23	20	19	233	16	41	12	9

Franklin and the smaller research submarines such as the US Navy NR1, the distinction between submarines and submersibles becomes less easy to define. The issue has therefore been side-stepped by calling this chapter Underwater Vehicles.

Underwater vehicles can range from the nuclear submarine carrying over 100 people including a restaurant, theatre, repair shops and a massive supply of nuclear weapons, to the tiny swimmer propulsion unit. To cover the range we may further divide them into two groups manned or unmanned, and these may be tethered i.e. powered by an umbilical or untethered (i.e. free swimming).

Manned Underwater Vehicles

1. *Nuclear submarine.* This vessel has excellent speed, depth and endurance capabilities with crews from 90 to 140 men. Range and speed match or exceed those of the conventional surface ship. See Fig. 4-1 for nuclear-sub details.
2. *Dry submersible.* This is generally a small low speed vessel carrying a crew of two to seven men at atmospheric pressure for short duration missions e.g. seabed survey, sampling, marine biological survey.
3. *Personnel transfer submersible.* The principal feature of this submersible is that it has a hatch which permits it to mate with another vehicle or base.
4. *Diver lock-out submersible.* The main feature of this submersible is to be able to transfer a diver at the work site from the submersible or vice versa.
5. *Swimmer vehicle or wet submersible.* This is a flooded submersible in which the diver is shielded from the water. Pas-

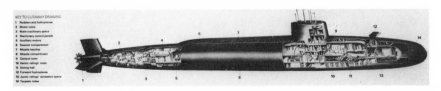

Fig. 4-1. How a nuclear submarine is put together.

sengers and crew may number from two to ten and breathe using their own or built-in breathing system (B.I.B.S.).

6. *Swimmer propulsion unit.*   Typically one man is either sitting astride or being towed.

Table 4-2 gives a list of larger submersibles, showing operator, home port, operating depth, area of operations, and nature of the work.

Existing manned submersibles are limited by the short endurance of the propulsion and auxiliary machinery. The U.S. Navy's 140 ft long NR1 (Fig. 4-2) is the only submersible with nuclear power to be built to date. The main power plant is a pressurized-water reactor generating steam for conventional turbine machinery. Built by General Dynamics Corporation and launched in 1969, the vessel has a beam of 12 ft and displaces 400 tons.

The crew numbers five with accommodation available for two passengers. The NR1 has an endurance limited only by the amount of food and supplies carried and crew fatigue. Main propulsion is via two propellers with four ducted propellers for maneuvering.

With its depth and endurance capabilities the NR1 will be able to perform tasks such as detailed mapping of the sea floor for scientific and commercial purposes.

A nuclear submarine tanker has been proposed, the initial design being a vessel 900 ft long and displacing over a quarter of a million tons when fully loaded. The route for this tanker is the NW passage or the Arctic Ocean under the ice, carrying as much as 170,000 tons of Alaskan crude oil from the North Slope to the East Coast of the United States.

Dry submersible.   This category of underwater vehicle probably includes the greatest range of vehicle size from the two-man Pisces "mini-sub" to the 130-ton Ben Franklin.

The Pisces submersibles are manufactured by HYCO of Vancouver. Eight have been built to date of which four are operated by Vickers Oceanics. Individual specifications vary, but Pisces III is typical of the class with a displacement of 11½ tons and an overall length of 20 ft. Maximum depth capability is 3,600 ft. See Figs. 4-3 and 4-4.

Vickers Oceanics operate the submersibles as part of a com-

TABLE 4-2
Larger Manned Submersibles

Submersible	Operator/Home Port	Operating Depth (Ft)	Area of Operations	Nature of Work
CANADA				
Aquarius I	P & O Intersubs, Vancouver, B.C.	1,200	Eastern Canada	Offshore Oil
Auguste Piccard	Horton Maritime Explorations Vancouver, B.C.	2,000	Undergoing modifications in Vancouver	Geophysical/Seismic surveys
Pisces IV	Department of Environment Vancouver, B.C.	6,500	Western Canada	Environmental Studies
Pisces V	Department of Environment Vancouver, B.C.	6,500	Eastern Canada	Cable Surveys/Burial
Pisces VII	P & O Intersubs Vancouver, B.C.	6,500	Under construction	Offshore oil/Cable burial
SDL-1	Canadian Armed Forces Halifax, Nova Scotia	2,000	Canadian waters	Military Tasks
Sea Otter	Arctic Marine Vancouver, B.C.	1,500	Western Canada	Cable and hardware surveys and inspection
FRANCE				
Archimede	French Navy Toulon	3,600	Mid-Atlantic Ridge	Scientific Research
Cyana	Centre National pour l'exploitation des Oceans, Paris	9,842	Mid-Atlantic Ridge	Scientific Research
Diving Saucer	Campagnes Oceanographique Francaises/Monaco	1,350	Worldwide	Photography
Griffon	French Navy Toulon	1,968	Presently under test and evaluation	Military Tasks

Name	Operator/Location		Area	Purpose
FRANCE cont.				
Sea Flea (2@)	Campagnes Oceanographique Francaises/Monaco	1,620	Presently not working	Photography
Shelf Diver	Inter-Sub Marseille	800	European waters/ Mediterranean	Offshore Oil/General engineering
ITALY				
PC-5C	Sub Sea Oil Services (S.P.A.)	1,200	Undergoing refit	Offshore Oil
PS-2	Sub Sea Oil Services (S.P.A.) Milan	1,025	North Sea	Offshore Oil
Tours 66	Sarda Estracione Lavorazione Cagliani	330	Mediterranean	Coral collection
JAPAN				
Hakuyo	Ocean Systems Japan Tokyo	984	Indo-Pacific	Fisheries Research
Kurushio II	University of Hokkaido Hokkaido	650	Japanese Waters	Fisheries Research
Shinkai	Japanese Maritime Safety Agency, Tokyo	1,968	Western Pacific	Scientific research
Yomuiri	Yomuiri Shimbu Newspaper Tokyo	792	Western Pacific	Scientific research
NETHERLANDS				
Nereid 330	Nereid NV, Schiedam	330	Unknown	Offshore Oil
Nereid 700	Nereid NV, Schiedam	700	Construction to be completed in 1974	
TAIWAN				
Tours 64	Kuofend Ocean Development Corp. Taipei	330	Western Pacific	Scientific Research

Larger Manned Submersibles

Submersible	Operator/Home Port	Operating Depth (Ft)	Area of Operations	Nature of Work
UNITED KINGDOM				
Moana	Comex Marseille	1,000	North Sea	Offshore Oil/Underwater Inspection Maintenance
PC-8	Northern Offshore Limited London	800	North Sea (assumed)	Offshore Oil
Pisces I	Vickers Oceanics Barrow-in-Furnace	1,800	North Sea	Offshore Oil
Pisces II	Vickers Oceanics Barrow-in-Furnace	3,000	North Sea	Offshore Oil
Pisces III	Vickers Oceanics Barrow-in-Furnace	3,000	North Sea	Offshore Oil
Pisces VIII	Vickers Oceanics Barrow-in-Furnace	3,300	Under Construction	
Taurus	P & O Intersubs Montrose	2,000	Under Construction	Offshore Oil
Vol LI	Vickers Oceanics Barrow-in-Furnace	1,200	North Sea	Offshore Oil
UNITED STATES				
Alvin	Woods Hole Oceanography Institute, Woods Hole, Mass.	12,000	North Atlantic	Scientific research
Asherah	Technoceans New York City	6,000	Presently not working	Scientific Research
Beaver	International Underwater Cont. New York City	2,000	North Atlantic	Offshore oil/cables

UNITED STATES
cont.

Deep Diver	Smithsonian Institute Ft. Pierce, Fla.	1,300	Decommissioned	General Scientific/ engineering
Deep Quest	Lockheed Missiles and Space Corp. San Diego, Calif.	8,000	Northeastern Pacific	General Scientific
Deepstar-2000	Western Corp. Annapolis. Md.	2,000	Presently not working	
DSRV 1	U.S. Navy San Diego, Calif.	5,000	Undergoing test and evaluation	Rescue
DSRV 2	U.S. Navy San Diego, Calif.	5,000	Undergoing test and evaluation	Rescue
Johnson-Sea-Link	Smithsonian Institute Ft. Pierce, Fla.	2,000	Bahamas/Florida	Scientific Research
Johnson-Sea-Link	Smithsonian Institute Ft. Pierce, Fla.	2,000	Construction to be com- pleted in 1974	
Nekton Alpha	General Oceanographics Newport Beach, Calif.	1,000	U.S. Coastal waters	Offshore oil/general scientific and hardware inspection

Fig. 4-2. The Navy's U.S. first nuclear-powered oceanographic research submarine.

prehensive underwater work service, comprising a support vessel with work shop and laboratories, rough weather handling system, complex telemetry and navigational aids as well as the submersibles themselves.

These in turn can be equipped with a variety of underwater tools—manipulators, torpedo grabs, rotary power tools, jetting nozzles—and for inspection work, closed circuit TV cameras with videotape recording facilities and thallium iodide lighting.

Typical tasks are torpedo recovery, oceanographical and geophysical work, pipeline survey, and cable burial.

Alvin. Slightly larger than the Pisces submersibles and used for generally different purposes, is Alvin, a deep submergence research vehicle owned by the U.S. Navy and operated by Woods

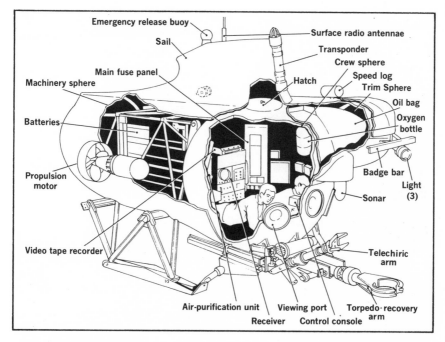

Fig. 4-3. The Pisces III submersible has displacement of 11½ tons, overall length of 20 ft.

Hole Oceanographic Institution, Fig. 4-5. The vessel is 23 ft long and displaces 16 tons.

In 1973, the depth capability was increased from 6000 ft to 11500 ft by the fitting of a new titanium pressure hull. 1968 saw the accidental sinking of Alvin in 5000 ft of water, but she was successfully rescued a year later with the assistance of the Aluminaut.

Hakuyo.   Another typical undersea work boat is the Hakuyo, Fig. 4-6. This 6.6-ton submersible was built by Kawasaki Heavy Industries and delivered to its owners, Ocean Systems Japan Limited, in 1971. Hakuyo has a depth capability of 980 ft and an endurance of five hours. Life support of 48 hours for three men is provided.

There is one main drive motor and three maneuvering thrust-

Fig. 4-4. Pisces V is one of the latest of the type built by HYCO.

ers. Vision is good with fourteen 6-inch diameter viewports. Other "standard" items are manipulator, sonar, and underwater telephone.

Ben Franklin.   At the top end of the submersible range is the Ben Franklin, Fig. 4-7, the largest commercially owned submersible in operation, displacing 130 tons with an overall length of 49 ft. The vessel was designed by Jacques Piccard and built by Grumman Aerospace Corporation who are also the owners.

Ben Franklin's first mission was to study the Gulf Stream and in 1969 the submersible drifted unpowered for one month from Florida to Nova Scotia (1500 miles) with a crew of six. Maximum operational depth on the voyage was 2000 ft.

Aluminaut.   The combat submarine has for many years incorporated ring stiffened cylinders in its pressure hull structure. More recently, the Reynolds Metals Company sponsored the construction of a deep submersible called Aluminaut, Fig. 4-8, utilizing very large stiffened cylindrical sections of 7079-T9 aluminum alloy.

66

Fig. 4-5. Submersible Alvin. Personnel sphere can be separated from the rest of the vessel. Courtesy Woods Hole Oceanographic Institution.

The Aluminaut is 51 ft, 3 in. long with an extreme beam of 15 ft, 4 in. and a height of 14 ft, 3 in. The design operating depth is 15,000 ft.

Making the pressure hull involved several new ideas in working with very large aluminum ingots and forgings. The cylindrical portion of the hull is composed of 11 identical flanged cylinders 97 in. OD x 84 in. ID x 40 in. long giving a wall thickness of 6.5 in. The flanges are 1.5 in. thick and 5 in. deep. Because 7079 alloy is not easy to weld, a bolted construction was adopted using about 400 anodized aluminum shrink-fit bolts. The holes were jig drilled and reamed in place to give an interface.

Opsub. Many submersibles are limited in operation by the capacity of their batteries. One way of improving the situation is to provide power to the submersible through an umbilical from the surface. Such an arrangement is used by Opsub, an 18-ft long, 4.6 ton vehicle.

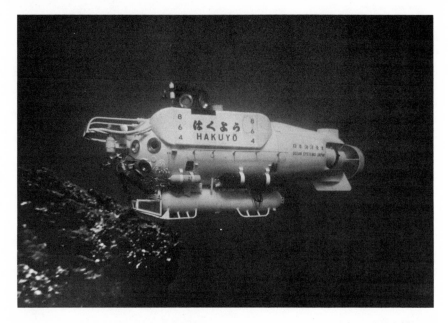

Fig. 4-6. The Hakuyo is a 6.6-ton submersible built by Kawasaki Heavy Industries for Ocean Systems Japan Ltd.

Opsub's endurance is limited only by its life support system which provides for its 2-man crew for 24 hours, with an extra two hours emergency breathing supply. The umbilical also carries the telephone and TV transmission lines to the surface. Opsub can presently operate to 1,000 ft water depth, though the pressure hull itself is designed for 2,000 ft.

Gvidon. The Gvidon, Fig. 4-9, is a Russian ship-attended two-man submarine, capable of diving to 820 ft. The hull is vertically orientated and houses control systems, scientific equipment and a regeneration plant for its crew.

It withstood rigorous launching lifting and maneuvering programs during tests in the Black Sea. Designed by engineers Igor Danilov and Oleg Pavlov of the USSR All-Union Research Institute of Marine Fish Economy and Oceanography, the vehicle is designed for mobile marine life observations, and serves as a habitat for inspecting seafloor objects and fishing equipment.

Fig. 4-7. The Ben Franklin, displacing 130 tons, is the largest commercially owned submersible in operation.

The vertically oriented hull houses control systems, scientific equipment, a regeneration plant, and two to four crewmen. Pilot remains in the lower section and the observer in the upper area for long-term observations. A four-man crew, consisting of a pilot and three observers, mans the submarine during periods not exceeding six hours.

Surface displacement of the Gvidon is 4 tons, with a submerged displacement of 4¼ tons. Vehicle measures 14.6 ft high, 8.2 ft long, and 8 ft wide. Emergency ballast adjusts from 660 to 700 lb.

The submarine is self-contained with power supplied by storage batteries rated for 400 ampere-hour performance. Electric power supply permits submarine to cruise for 5 hours, and to regenerate air for 72 hours.

Fig. 4-8. Reynolds Metals Co.'s Aluminaut utilizes very large cylindrical sections of a special aluminum alloy.

Vehicle is not designed for observations of active fishing gear like trawls. Therefore, high speeds are not required. The Gvidon will travel from 0.5 to 1.5 knots, which is satisfactory for conducting observations required.

High maneuverability in the horizontal plane is due to low speeds and unit design. Propulsion devices, which run clockwise and counter-clockwise, enable the vehicle to make a complete revolution in several seconds. The Gvidon stops almost at once when both propellers are disengaged.

Observations in dark layers of water are facilitated by lights mounted on side of the shell. Lighting system consists of 9 units, with 2 in the nose of the propellers, 3 on the support elements, 2 in the vehicle's mid-section, 1 at the entrance hatch, and 1 sliding light on a downward oriented telescopic rod.

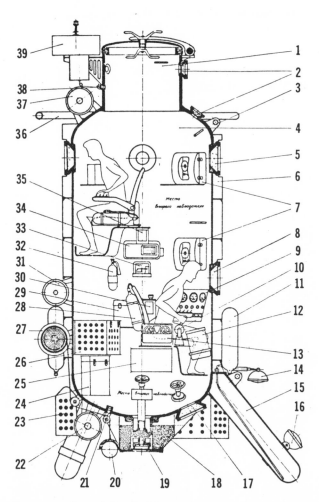

Fig. 4-9. Cross section of the Gvidon submarine research vehicle showing (1) entrance hatch, (2) portholes, (3) lifter bracket, (4) robust hull, (5) observer's porthole, (6) ballast tank, (7) fathometer, (8) pilot's porthole, (9) electric panel, (10) propeller turn handle, (11) high pressure air tanks, (12) directional gyro (13) dictaphone, (14) telescopic lamp, (15) equalizing tank, (16) light, (17) bottom porthole, (18) emergency ballast, (19) ballast release mechanism, (20) guide rope, (21) guide rope release mechanism, (22) electric cam case, (23) storage batteries, (24) regeneration element storage, (25) regeneration element, (26) regeneration plant, (27) guide rope winch, (28) emergency respiration apparatus, (29) gas analyzer for oxygen, (30) gas analyzer for carbon dioxide, (31) electric cam case, (32) fire extinguisher, (33) barometer, (34) radio station, (35) emergency battery, (36) safety arch, (37) electric cam case, (38) emergency buoy release mechanism, and (39) emergency buoy.

Personnel Transfer Submersible

The need to carry out deep sea rescue and recovery became apparent with the loss of the nuclear submarine Thresher in April 1963, and the loss of a hydrogen bomb off the coast of Spain in 1966.

Thresher was lost in about 8,490 ft, 250 miles off the coast of Boston. The deep diving bathyscaphe Tieste II was towed to site, and with side scan sonar, quartz iodine lights, electronic strobe and underwater cameras, located and photographed what remained.

Because of this tragic loss the Deep Submergence Systems Project was established in June 1964. The Project covers five areas:

1. Submarine location, escape and rescue.
2. Object location and small-object recovery.
3. Large-object salvage system.
4. Nuclear-powered, deep submergence, Research and Ocean Engineering Vehicle NR-1.
5. Man-in-the-Sea (The work on this was mainly covered by the SEALAB III development.)

The major effort in this program has been the development of the Deep Submergence Rescue Vehicle (DSRV). Its principal characteristics are:

Overall length	49 ft, 3 in.
Diameter	8 ft
Propulsion	Electric motors, batteries
Maximum speed	5 knots
Endurance on bottom	3 knots for 12 hours
Operating depth	5,000 ft
Crew	Pilot, copilot
Passengers	24
Life support	24 hours minimum plus individual life support for 3 hours minimum

The outer hull is constructed of fiber glass, and within this are three interconnected spheres which form the manned pressure capsule.

Each sphere is 7½ ft in diameter and is constructed of HY 140 steel. The forward sphere contains the vehicles control equipment and is manned by a pilot and copilot. The center and aft spheres accommodate 24 passengers and a third crewman.

Submarine escape. The DSRV provides an external means of submarine rescue. There is a need, however, to permit unaided escape from a stricken submarine.

Submarine escape is done in two stages. First, the escaping personnel must pressurize inside the submarine from the internal pressure (probably close to atmospheric) to the ambient water pressure; and secondly, they must exit the vessel and rise to the surface, finishing at one atmosphere again.

To minimize the adverse physiological effects of high pressure, it is desirable to achieve the first stage as rapidly as possible. This is best done by using a one-man escape tower from which to make the ascent. To speed the process even more, the tower is flooded prior to pressurizing. This necessitates the use of a special Submarine Escape Immersion Suit, a water-tight garment connected by quick release coupling to the Built-In breathing System (BIBS). See Fig. 4-10.

Impurities considered minor and breathable at one atmosphere can be fatal under pressure. BIBS therefore supplies a very pure supply of high pressure air.

An escape from 500 feet can be completed in eighty-five seconds, thus removing physiological complications.

Diver lockout. The diver lockout is a combination of dry submersible and diving bell. The vessel is divided into two major compartments, one permanently at atmospheric pressure and manned by the vehicle's crew, and the other pressurized to the working depth and accommodating the saturation divers.

The Perry PC 1202, Figs. 4-11 and 4-12, built for Vickers, is a 13-ton diver-lockout submersible capable of operating down to 1,200 feet. It has a length of 32 ft and features a transparent acrylic window providing unrestricted viewing ahead. The vehicle has a two-man crew and a breathing supply for two divers. In addition to work performed by the divers, it can perform tasks using its own manipulators.

On retrieval by its mothership, PC1202 is mated to a decompression chamber and the saturation divers are transferred. The

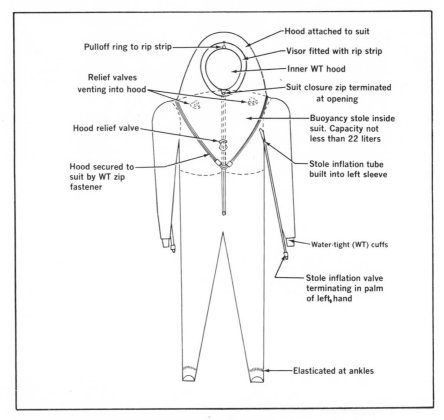

Fig. 4-10. Submarine-escape immersion suit.

lockout can then, if desired, be relaunched with a new team of divers.

Swimmer vehicles (wet submersibles). A wet submersible, as the name implies, has a flooded cockpit (at ambient pressure) as opposed to the dry submersible's atmospheric environment. The main advantage is that no pressure sphere is required, which realizes a large saving in capital cost.

The wet submersible is similar in certain ways to the diver lockout. In both cases divers can make sorties from their vessel, using it as an underwater workshop for rotary, cutting and welding equipment. The wet submersible cannot however be used for saturation diving though mixed gas breathing is feasible.

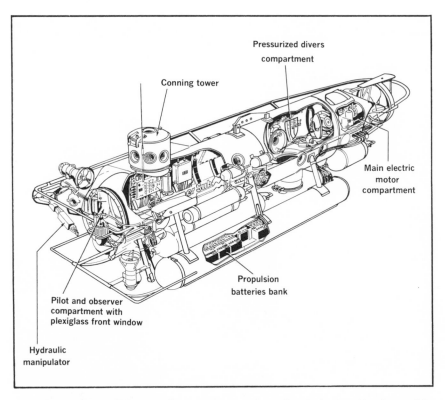

Labels on figure:
- Pressurized divers compartment
- Conning tower
- Main electric motor compartment
- Propulsion batteries bank
- Pilot and observer compartment with plexiglass front window
- Hydraulic manipulator

Fig. 4-11. Perry PC1202 diver-lockout submarine built for Northern Off-shore Ltd. and InterSub Ltd.

The Cooke Bros. DT V2 Submersible, Fig. 4-13, is a two-man vehicle with a cruising range of 25 miles at 5 knots. At 1 knot it has an endurance of 30 hours on one battery charge. Depth capability of the vehicle alone is 180 m, but operational depth is of course dependent on the need for decompression stops.

Swimmer Propulsion Unit (SPU). A modern one-man torpedo-shaped SPU is the Dimitry Rebikoff "Pegasus." This vehicle has a 10 in. diameter x 10 ft, 6 in. body with retractable diving planes near the bow. Extended, they give the vehicle a beam of 47 in. Different models of Pegasus have been produced with maximum speeds varying between three to six knots, and ranges from five to ten miles.

Fig. 4-12. Diver and submarine work as a team in the PC1202 built by Perry Oceanographics, Inc.

Fig. 4-13. Cooke Bros. DTV2 submersible is a two-man vehicle.

The underwater duration of the vehicle on one battery charge is 1½ hours. Recharging is rapid enough to allow two dives per day down to a maximum of 250 ft. Pegasus weight in air is 290 lb.

Pegasus is used primarily for underwater survey work with a variety of cameras (still, cine, and videotape) mounted at the forward end. Associated lighting is also provided, enabling high quality photography in the worst conditions.

Safety of manned systems.   First, the importance of having a proper mother ship can hardly be overstressed.

The percentage of time lost due to weather, the safety and overall efficiency of the operations depend to a large extent on the suitability of the mother ship. The size, design and fitting-out of the mother ship (or surface support unit) will be determined by the size, design, range and seaworthiness of the surfaced underwater vehicle.

On remote missions mother ship would need to carry two vehicles. It must be noted that the vehicle not only has to withstand great pressure at depth but also resist heavy wave forces on the surface hence seaworthiness remains a vital factor.

The pressure hull visually fixes both buoyancy and the weight of the vehicle. If the weight is much greater than the buoyancy, the design is not viable.

The margin by which buoyancy exceeds the weight represents the second critical factor—the payload. The payload margin can be improved just as in an aircraft by incorporating light or strong materials in the design.

Once the designer has arrived at a vehicle with required performance, consideration must be given to the safety of the operators and crew.

It must be realized that no component or personnel can be considered to be absolutely safe, i.e. 100%. Additional fail-safe systems are therefore introduced, but at the expense of the payload. The number of systems have to be equated to the number and type of people at risk. Comparison can be made between the systems of a nuclear submarine and a dry submersible. A dry submersible must operate with the utmost reliability, manned by one to two well-trained men. The risk is then as low as a small executive jet on a scheduled flight.

Several forms of hazard can be envisaged.
1. Uncontrolled leakage into the crew's compartment while submerged. This would be fatal.
2. Uncontrolled leakage into an unmanned compartment. The vehicle would probably sink but the crew could survive, providing their compartment was not crushed by the increased pressure.
3. The vehicle becoming entangled with something on the seabed from which the crew cannot release it.
4. Electrical power failure, preventing operation of valves external to the pressure hull and the controlled-buoyancy system.
5. Rapid failure of the life support system, particularly the

oxygen in the control compartment which would lead to a loss of control.

During offshore operations, it is essential for the surface vessel to maintain continuous communication with the vehicle; thus, if an emergency occurs there is no delay in effecting recovery. If communications break down, it is possible in an emergency to release a tethered buoy.

Most vehicles jettison ballast, tools, and other items to increase buoyancy. However, if serious flooding or entanglement occurs this may not be sufficient to return the vehicle to the surface.

Recoveries to the surface may therefore be carried out by the use of another vehicle (manned or unmanned) which can attach lifting lines.

Future possibilities would be the design of supplementary buoyancy packs, something which could be inflated underwater. Compressed air would be impractical, but a chemical product such as lithium hydride might be used. When brought into contact with salt water, it generates hydrogen, but as yet, the effects of considerable pressure on it are not known. Another possibility would be to release the crew compartment; however, this release must be quick, clean, and certain. To safeguard the crew, this recoverable compartment would need to contain a complete life-support unit. So far, it has been preferred to recover the complete vehicle.

The life-support systems will determine the time available for the rescue operations. On the Vickers submersibles the life-support system now provides for 320 man hours (160 crew hours).

As a general rule, the life-support should be determined by the time it would take to complete the rescue, making a fair allowance for snags. Not only the vehicle, but the entire operation must be considered. During a rescue, the demands on the mother ships would be different from normal operations.

Unmanned Underwater Vehicles

Torpedos (untethered). This category is based on the military torpedo with explosive warhead. A first development was the mobile

79

target which, instead of carrying a warhead, contained navigation and sonar equipment for it to simulate submarine behavior and echo propagation. A number of such vehicles have been made by Recording Designs Ltd. (RDL).

These target torpedos can be adapted for underwater survey work but certain modifications are desirable. First, as targets, endurance is sacrificed for speed. As survey vehicles, a slow speed, long endurance configuration is preferable, necessitating a change in the propulsion motor and gearing.

Another feature of the mobile target is its low displacement. If the dimensions are increased, greater payloads can be borne, permitting more sophisticated survey equipment to be used. The University of Washington's SPURU possesses these improvements.

CURV Cable controlled underwater recovery vehicle (tethered). It was CURV in conjunction with Alvin, a two-man research submarine and Aluminaut which recovered the lost bomb in the waters off Palomares Spain in 1966.

In 1973, the CURV assisted in the rescue of the Pisces III which had sunk in 1,530 ft, 250 miles off Cork, Eire. In this operation CURV carried a heavy lift line in its manipulator down and attached it to the Pisces.

The structure and configuration of CURV are intentionally simple and non-hydrodynamically designed to afford maximum adaptability and versatility. The basic system consists of an open aluminum rectangular frame to which support systems can be readily and efficiently added to modify the vehicle for a required task. See Fig. 4-14.

The frame comprises the body of the vehicle; the other components of the basic system include the control cable, control console in a portable van, power supply and conversion equipment, surface handling equipment, and the YFNX-30 support vessel equipped with a Boat-Mounted Acoustic Locating Device (BALD). Syntactic foam blocks attached at the top of the frame produce a slight positive buoyancy.

The support systems which are mounted on the frame include: optics, sonar, propulsion, hydraulics, compass, and tool assembly. The actual equipment involves active and passive sonar, two closed-

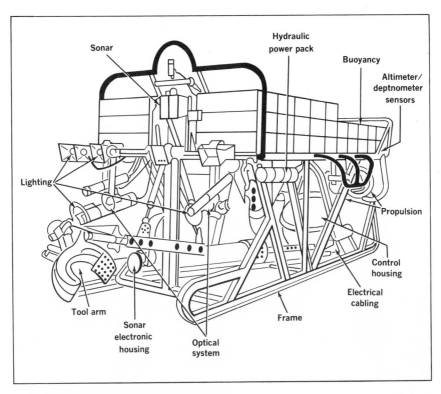

Fig. 4-14. CURV IIIC vehicle subsystem.

circuit television cameras, a 35-mm camera, underwater lights, three 10-hp propulsion motors, and an electrohydraulic command system which includes the claw.

Nominal operating depth is 7,000 ft; emergency mission capability to 10,000 ft. CURV is a proven system for inspection, recovery, work tasks, and small area (1-2 square mile) search missions.

It has a normal crew of seven and emergency mission crew of ten.

The support-ship requirements include station keeping and cable handling area away from crews. Deck space for approximately 26 short tons of equipment with a total cubic size of 4,500 ct ft (seven items approximately 75-120 sq ft each); plus handling crane for vehicle.

Fig. 4-15.   Kematsu amphibious dozer.

Fig. 4-16. The Hitachi robot.

CURV is capable of load-out in one C-141 for shallow missions (to depths of 1,500 ft) on known ship of opportunity; or two C-141 for unknown support ship of emergency deep-ocean missions.

Other vessels which can work on the seabed include Kematsu's amphibious dozer (Fig. 4-15) and the Hitachi robot (Fig. 4-16).

Underwater trencher. A novel underwater trencher has been developed by Sumitomo Shipbuilding and Machinery Co. Ltd., along with a number of other Japanese firms. The trencher, Fig. 4-17, is unmanned and consists of a cutter-suction dredger mounted on a tracked chassis. The vehicle is controlled via an umbilical from the surface.

Both control and traction systems are electro-hydraulic, the latter giving the vehicle a top speed of 2 mph. The trencher has a pump motor of 75 kw, suitable for sand and clay. The vehicle's dimensions are 11.7m x 5m.

An important feature of the trencher is that all its components are modularized for rapid maintenance.

Fig. 4-17. Underwater trencher built by Sumitomo Shipbuilding and Machinery Co. Ltd.

The advantage of the trencher over a conventional surface suction-cutter dredger is that it is unaffected by adverse waves or currents.

Pipe Burying Machine

The Pipe Burying Machine (PBM) is a seabed vehicle for trenching under and burying submarine pipelines. See Fig. 4-18.

The PBM is installed and recovered from the pipeline using controlled buoyancy guide lines laid by divers and down hand winches, trenching operations can then commence. Progress is maintained by a push-pull system along the line.

There is also a manual override if at any time the operation needs to be commanded from the surface or seabed. This method

Fig. 4-18. Pipe-burying machine of Sub Sea Oil Services trenches under and buries submarine pipeline.

has the advantages of providing an automatically controlled force, and reliable endurance. The method also imposes less strain on the structure of the pipe itself (the maximum force acts along the center line of the pipe being 5 tons).

The use of two quick-release weak-link joints and a constant tension on the umbilical line from the PBM to the vessel also ensures that the stresses imposed on the line are within design limits.

The controls can be set to produce any trench width between 1.8 and 4.4 meters and an initial depth of 1½ to 2 meters. The seabed itself will have a direct effect on these operating parameters and these are determined by the torque and speed of the cutters. This velocity, force and torque are directly measured and recorded onboard the surface support vessel. It is also possible to measure the density of the slurry and monitor the forces on the push-pull together with the rate of progress.

Inspection is carried out by closed-circuit TV and the discharge system has been arranged not to affect the visibility.

5

Underwater Power
Sources

TO carry out work underwater and to extend the range and performance of the diver and vehicles, power is required. There is wide use of power and many systems have been and are continuing to be developed.

They cover ordinary batteries for hand-held torches, large battery packs for submersibles, and pneumatic and hydraulic systems for tools to nuclear power for submarines.

These forms are available: (1) manual (the diver), (2) dynamic/mechanical, (3) pneumatic, (4) hydraulic, (5) electrical, and (6) nuclear.

The manual power of divers alone is limited, particularly in cold waters, and the addition of simple mechanical aids such as levers, hammers, and spanners can improve his efforts.

Pneumatic and hydraulic are especially used for hand-held tools and for this purpose are extremely effective, although pneumatic power is limited to a depth of about 50 meters, due to the practical problem of hose handling and the power required to overcome the static head (depth/pressure).

Hydraulic power (in the form of accumulators) has a wider use but is often associated with electrical power for control systems.

It is perhaps not surprising that electrical power (despite the apparent disadvantages of the watery environment) has proved to be the most versatile form of power. It is in fact not the generation of the electrical energy that presents the main problems but distribution and in particular connections.

Nuclear reactors are suitable for high power levels, generally in excess of 100 kw which can be sustained over a period of months. At present they are used underwater only in the military application for submarines.

The future may well require an extension of the nuclear power as used in submarines to offshore and seabed power stations.

Manual

The diver has only his own muscle power to move things around and this manual effort is severely limited by cold water and with increasing depth (pressure) his performance also becomes impaired.

There is a need to develop purpose designed hand tools for the diver, known affectionately as "handraulic" tools—tools with springs, static hydraulics, or other means of obtaining maximum mechanical advantages for the diver with the minimum preparation, time and encumbrance from umbilical hoses. The deep diver in particular is likely to have umbilical problems just getting himself to the work site.

An average "free-swimming" diver is unlikely to be able to maintain a speed of more than ½ knot for a period longer than ½ hour (compare manned submersibles with operating speeds of 1½ knots, and duration of a mission lasting 4 to 6 hours).

Dynamic/Mechanical

Apart from simple mechanical devices, e.g. levers and hammers, there are various forms of dynamic energy converters which can be used underwater. These fall into four main categories.

1. *Open cycle, internal-combustion engines.* In this cycle the fuel is burned in the engine inlet gas stream so that the products of combustion form part of the working fluid. Examples are diesel, petrol and gas turbine engines.

2. *Closed or partially closed cycle, internal combustion engines.* All engines in the first category can in theory be operated in the

closed cycle, i.e., part of the exhaust recycled after fresh oxidant added.

The only practical rivals to a battery system or fuel cell as the prime power source for civilian submersibles are engines using vapor or gas cycles, such as the closed cycle diesel. But many of these suffer from disadvantages, such as exhaust back pressure, disposal of CO_2, and noise.

Engines of this type were developed for shallow diving military submarines by the German Navy towards the end of World War II. This work has recently been revised in the U.K. with a diesel engine which will generate on the surface in a conventional manner or submerged in a sealed container drawing its oxidant from a stored supply.

Numerous thermal systems are possible for incorporation as the propelling agent in small submersibles. Many variations of vapor and gas cycles have been proposed, some of which have been built and tried. Current interest in the recycle diesel is a revival of work established several years ago. In World War II the German Navy developed and introduced a recycle diesel using hydrogen peroxide (HTP) as the oxidant.

They also developed a system for miniature military submersibles, but the introduction was forestalled by the end of hostilities. After the war the Royal Navy did considerable development work on this type of system until 1958 when all work stopped as a result of the introduction of nuclear propulsion.

In the U.S.A. development followed similar lines to that in the U.K., culminating in the building by Fairchild of a shallow-diving, three-man vessel propelled by a closed cycle diesel using HTP as the oxidant.

A second thermal system was also introduced by the German Navy, the Walter turbine, of which there were two types. The "indirect" one was a conventional closed stream turbine fed from a pressurized boiler using HTP as the fuel. The "direct" system was a steam/gas turbine operating on combustion products of fuel in decomposed HTP, diluted and recirculated water. Two high-speed target intermediate submarines were built by the RN after World War II using the direct system.

3. *Externally heated closed cycle.* This category covers the Rankine and closed cycle gas turbines and the Stirling engine.

4. *Mechanical system.* This class comprises dynamic power sources which are not based upon fuel burning engines but in which power is derived from some form of stored or locally available mechanized energy e.g. hydrostatic.

Pneumatic

The pneumatic tools used by divers are essentially the same as the tools used on the surface. There is a depth limit of about 50 meters, in sizing the compressor for open circuit tools, due to pressure loss along the hoses and the increased pressure resistance due to depth.

Closed circuit pneumatic tools could be used, but the tools would then need to withstand the full external water pressure less the internal line pressure. Also an additional hose would be required which would be cumbersome for the divers to handle.

Hydraulic

These tools utilize the flow of relatively incompressible fluids. They are normally operated in the closed circuit mode and therefore two hoses are used, which does impose a burden on the diver.

Hydraulic control systems are used in offshore drilling on valves and chokes, and in submersibles.

For deep diving tasks, beyond 200 meters, it is most convenient to use pressure-balanced power-tool systems, which means using a working fluid capable of coping with pressure one way or another. Work has also been carried out on using filtered sea water as the fluid.

Other developments are pressure balanced electro-hydraulic power using conventional hydraulic principles and submerging the pump and motor in the working fluid. The power for the unit is supplied by an electrical umbilical from the surface. When possible, tools will be fitted with concentric hydraulic hoses to reduce the negative buoyancy effects.

These tools unfortunately are heavy and need buoyancy compensation. However, reliable suction pads have been developed to assist in bracing the diver against his work force.

Electrical

With very few exceptions all the current civilian submersibles, manned and unmanned, are propelled by electrical power or electro-hydraulic systems. The two exceptions are the "TOURS" design from West Germany which can include a conventional diesel engine for passage on the surface and the "Deep Voyager" a design of the Makapuu Oceanic Center Hawaii, which makes way by gliding during alternate descending and ascending modes.

Table 5-1 lists some of the current manned submersibles and their power sources.

Electrical Batteries

Lead/acid batteries have been favored in the majority of manned submersibles, with silver zinc in four; nickel cadmium, two; and fuel cells, one. Other batteries which have been suggested are sea-water-activated magnesium-silver chloride, and magnesium-cuprous chloride, some of which are used in torpedoes where a very high discharge rate is required but their characteristics are totally unsuited to the slow discharge, long endurance required by a manned submersible.

The great attraction of the lead/acid system is that it is the cheapest off-the-shelf battery available but it pays for this with a low energy density. A total of 1500 charge/discharge cycles can be obtained with a discharge factor of 80% and an average discharge voltage of 1.9 v. Increased energy densities can be obtained at the expense of reducing the number of cycles.

Silver/zinc batteries are considerably more expensive than lead/acid, but give a higher energy density to to 90 Wh/kg. Their performance, however, is limited to approximately 150 cycles, with a discharge factor of 70% and an average discharge voltage of 1.5 v.

Nickel/cadmium batteries have lower energy densities than

TABLE 5-1

Power Sources for Submersibles

Submersible	Battery	Submersible	Battery
ALVIN	Lead/acid 87kwh	ALUMINAUT	Silver/zinc 190kwh
AMERICAN SUB CO.600	Nickel/cadmium	BEAVER	Lead/acid 30kwh
BEN FRANKLIN	Lead/acid 750kwh	BENTHOS V	Nickel/cadmium
DEEP STAR 4000	Lead/acid 43kwh	DEEP STAR 2000	Lead/acid 26.5kwh
DSRV I & II	Silver/zinc 56kwh	DEEP STAR 20000 (design only)	Silver/zinc 75kwh
HAKUYO	Lead/acid 72kwh	DSSV (design only)	Fuel cell 110kwh
PICES I	Lead/acid 60kwh	NEKTON ALPHA & NEKTON BETA	Lead/acid 4kwh
PLC4 (DEEP DIVER)	Lead/acid 34kwh	PC4B (SHELF DIVER)	Lead/acid 35kwh
TOURS 73	Lead/acid 50kwh	PICES II & III	Lead/acid 55kwh
		STAR III	Lead/acid 30kwh

lead/acid, the same discharge factor, and an average discharge voltage of 1.1 v; their big advantage is that as many as 4,000 charge/discharge cycles are obtainable.

A silver/cadmium battery has about half the energy density per unit weight or volume of a silver/zinc battery.

The operational significance of changing from lead/acid to silver/zinc or silver/cadmium batteries is shown in Table 5-2, the performance characteristics of the Perry PC3A design.

The great preponderance of lead/acid batteries in Table 5-2 gives weight to the criticism that the majority of the current range of civilian manned submersibles were designed with no specific task in mind other than to get into the ocean as quickly and cheaply as possible and then to think of some task that the vehicles could execute.

In designing a power system to meet specific operational requirements it should be remembered that the discharge factor, the charge rate, and the number of charge/discharge cycles will vary inversely with the energy density. Slow discharge means the highest energy density per kilogram of materials.

When engineering the system, the range of temperature likely to be encountered should be fully borne in mind. It is possible to carry out charging on the support ship's open deck, followed by plunging the batteries into near freezing water shortly afterwards. Battery capacity is very dependent on temperature.

Battery systems under development include zinc/air, zinc/oxygen, and lithium/nickel halide combinations but all of these are still some way from practical economic propositions.

While the lead battery is today's best compromise for use in civilian submersibles and other subsea operations, there is a need for higher power and energy, by a factor of ten or more where the size and weight penalty of lead acid batteries will become overwhelming. The supplementary power packs and the power modules must have high integrity and be simple rugged units.

It is therefore likely that iso-propylnitrate (IPN) many find an application as a power pack and that fuel cells might win a place to provide low power but the highest power requirements will await

TABLE 5-2
Battery Characteristics

	Speed		Time		Distance	
	Surface	Submerged	Surface	Submerged	Surface	Submerged
Lead/acid						
Full power	5 knots	4¼ knots	30 min	30 min	2½ min	2 min
Cruise	3½ "	3 "	3 h	3 h	10½ "	9 "
Slow	2 "	2 "	8 h	8 h	16 "	16 "
Silver/zinc						
Full power	5 knots	4¼ knots	4 h	4 h	20 min	17 min
Cruise	3½ "	3 "	8 "	8 "	28 "	24 "
Slow	2 "	2 "	20 "	20 "	40 "	40 "
Silver/cadmium						
Full power	5 knots	4¼ knots	2 h	2 h	10 min	8½ min
Cruise	3½ "	2 "	4 "	4 "	14 "	12 "
Slow	2 "	2 "	10 "	10 "	20 "	20 "

the arrival of nuclear power and then the era of an underwater power station.

Fuel Cell Systems

There have been three demonstrations of fuel cells in an underwater nole; the first was the hydrazine-oxygen system installed in Star I for a short time; the second was the small unit used in U.K. by the Imperial College/Enfield Technical College inflatable habitat off Malta; and finally the Pratt and Whitney fuel cell was used with the Perry Submarine Builders habitat for 48 hours in 15m of water off the coast of Florida.

Fuel cells are not new in concept, the principle first being demonstrated in 1842 by Sir William Grove. Chemical energy of conventional fuels is converted directly into low voltage direct current. In many ways the fuel cell resembles a conventional battery with the fundamental difference that a battery has a finite capacity limited by the amount of reactants which can be packed into the electrodes.

In the fuel cell the reactants are fed in continuously, that to the anode being referred to as the fuel and that to the cathode as the oxidant. Other components of the system are the acid or base electrolytes that act as the ion transport medium, a catalyst to promote reaction at the electrodes, and provision for the removal of heat and waste products.

Conversion efficiencies as high as 80% can be obtained from some of the various combinations of fuel, oxidant, and catalyst. Fuels which have been used include hydrogen (either cryogenic or generated from ammonia), hydrocarbons, methanol, hydrazine, alkali metals, and borohydrites. Oxygen is the commonest oxidant supplied in cryogenic form but some systems have used air or hydrogen peroxide.

The system which shows the most promise for manned civilian submersibles is the hydrogen/oxygen combination using cryogenically stored reactants. A comparison of a silver/zinc battery against a hydrogen/oxygen fuel cell is shown in Table 5-3.

Other electrical power sources. Batteries and fuel cells are the only current means of generating the energy required for manned

TABLE 5-3
Battery, Fuel-cell Comparison

Requirements........20-25kw average load
40-50kw peak load
Total Energy 1100kwh
Silver zinc battery........weight 10400 kg, volume 6.1 m^3
Fuel cell ....weight 4500 kg, volume 3.5 m^3

submersibles. Other electrical methods suggested have been thermionic generators and thermoelectric generators, but neither of these shows any great promise of producing the energy densities required.

Electrical control. Both ac and dc propulsion motors have been used in manned submersibles. The propulsion pods, usually external to the pressure hull, are subject to the full diving pressure and problems could arise with the armature seal but pressure balanced techniques are usually used on the motor case.

Squirrel-cage ac motors have the greatest constructional advantages because, with the lack of brush-gear, the motor housing can be completely filled with water and pressure balanced automatically. In the "Ben Franklin" the propulsion system consists of four three-phase motors drawing their supply from static invertors.

When a dc motor is used it must be completely pressure resistant or filled with oil and pressure balanced. A number of sophisticated control circuits are now available using the pulse space method of controlling the power, a method achieving a great saving in power over the old series resistor or carbon pile methods of speed control.

Subsea Production

A considerable direct experience and involvement with underwater power sources was gained on the Zakum Subsea Production Scheme. This scheme, operated in the Arabian Gulf, was the first attempt at complete subsea production. Included in the scheme were numerous power sources which were evaluated during the period of operation.

On this scheme, Alan Webb, a senior electrical engineer with

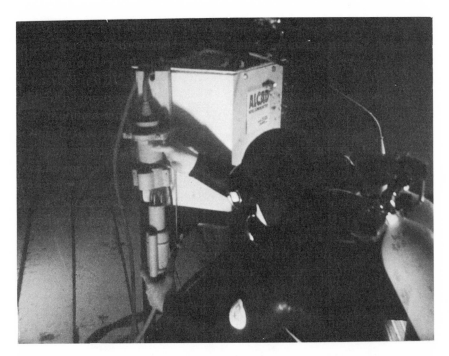

Fig. 5-1. Invention of BP's Alan Webb produces underwater electrical power by using oil-encased batteries and plug-and-socket joints.

BP, was the man responsible for power systems and distribution. See Figs. 5-1, 5-2, and 5-3.

As a result of the experience gained at Zakum some basic lessons were learned.

First, electricity and water do not mix although they can exist together. It is not such a shocking (if you will excuse the pun) combination and in fact there are advantages. The temperatures and climatic variations are minimal at great depths and a somewhat less obvious advantage, the equipment is out of reach of meddling fingers.

The circuits were developed around a power reservoir consisting of a large 24-v 600-amp-hr nickel cadmium battery of two 20-cell conventional 300-amp-hr batteries, housed in a seabed enclosure at atmospheric pressure and connected in parallel pairs (Fig. 5-2). This system was used so that after the trials the cells could

Fig. 5-2. Diver removing and unbolting a sub-sea battery enclosure, with a lifting parachute attached, in the Zakum field.

be incorporated directly into the above-water production telemetry system from Das Island.

To eliminate gassing problems the battery capacity was regarded as 300 amp-hr and terminal voltage was never allowed to exceed the gassing voltage of the electrolyte which, at the high ambient seawater temperature, was 1.37 v/cell.

All battery-charging connections included a voltage-sensing relay set to disconnect the charging supply when battery-terminal voltage approached 27.4v (20 × 1.37). The relays were arranged to have a wide differential of 4v to prevent hunting and to ensure that a degree of charge cycling was achieved.

The low charge rates and small loads applied to this massive battery meant that it was subjected to negligible duty. After 8 months on the seabed, gas evolution was undetectable.

The heavy battery enclosure was equipped with buoyancy tanks

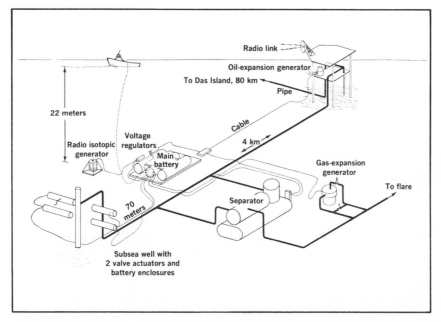

Fig. 5-3. Schematic diagram of Zakum sub-sea production-scheme electrical system.

to allow towing to and from the trial site. Control valves allowed divers to adjust the buoyancy while raising and lowering the enclosure to the seabed. A water-leakage alarm was included, as well as mercury-isolation switches operable by divers using a linkage sealed through a steel bellows. Provision was made for emergency charging through a shipborne generator.

Power Generation

The only form of submarine power generator available was a radio-isotopic thermoelectric type of 20-w output, obtained on loan from Snecma, the French Government agency.

Known as the Marguerite 20, this heavy generator was installed nearby and connected to charge the main battery with a voltage-sensing relay in the circuit. It worked to expectations until the output plug-and-socket connector failed.

Power to maintain the subsea site was derived from an in-line crude-oil expansion turbogenerator installed on a wellhead platform. This generator was one of some 30 units developed by J and S Pumps Ltd. to BP specifications to power wellhead telemetry and control. This drives a permanent-magnet three-phase generator to produce up to 600w of power at around 36v at 1800240Hz according to load conditions.

The generator terminals were connected to a variable transformer and then through a full-wave rectifier fitted with smoothing devices. The dc output was passed through the 4-km submarine cable directly to the main submarine battery through a voltage-sensing relay.

Battery voltage was indicated at the platform using a pair of conductors in the cable. By measuring dc current at the platform and adjusting the variable transformer, a steady charge of up to 4 amp could readily be supplied to the battery. The platform equipment was housed in Simplex-GE enclosures certified Division I flameproof to meet safety classifications.

It was suggested that a submarine version of the oil-expansion turbogenerator should be tested at the sub-sea site, but this was not considered useful, as the unit was glandless and designed to contain wellhead shut-in pressures and was already adequate for submarine use. However a gas-expansion turbogenerator by the same manufacturer was built as a seabed device for use on the gas line from the separators.

The machine was self-contained, the turbine being driven by a 40-psi pressure drop between machine inlet and outlet, maintained by a self-acting differential-pressure control valve. The turbine was a conventional Terry wheel-mounted on a vertical spindle driving an inductor alternator with an output of 40 kva at 100 v, 200 hz.

The alternator was designed to operate in a gas atmosphere at system pressure but included a unique bearing-lubrication system developed during experience on land. A further feature of this machine was a small permanent-magnet alternator fitted onto the main spindle extension to provide main generator excitation and to power a direct-acting, fully modulating, glandless 2-in. butterfly valve as speed controller in the gas-outlet pipe.

The unit included a 24-v charging circuit to provide power for the main battery through a voltage-sensing relay. This generator had run for many months on test but arrived too late for inclusion in the subsea trials.

Power Cable. The main cable between the battery and wellhead contained four conductors of 17 sq mm used in parallel pairs to carry the dc charging current. The conductors were insulated with PVC and screened with aluminum tapes. Around these were 44-1.5 sq mm PVC-insulated copper conductors for control indication and communications. The outer wrappings comprised copper tapes for marine-borer (Torpedo worms) protection, two layers of heavy steel armor wires and PVC sheathings. A lead sheath ensured the total cable was heavy enough to remain in place under sub-sea forces.

The 2½-in.-diameter cable was laid in a figure-eight format on a pontoon and the four drum lengths, weighing some 42 tons, were jointed before towing to the trial site.

This phase took some 2 weeks. Laying was accomplished in 1½ hr under the supervision of NKF of Delft, Holland, who designed the cable.

The submarine end-terminal box had been fitted at the factory and was lowered down a guide wire to an anchor pin on the main battery enclosure. The pontoon was then towed towards the wellhead platform. The tug's skipper received line-of-sight direction using a walkie-talkie on board a ship moored over the battery enclosure.

Biggest trouble. The main problem throughout was that plug-and-socket connections suffered total failure on a number of occasions. Eventually, modifications were made to an existing connector, and the result was a BP-patented submarine plug and socket. This device is oil-filled and can be made or broken underwater—it can even be left unconnected with the power on.

One prototype has been operated successfully in a permanent installation for 1 year at 100-v dc. The joint has been made and broken underwater 21 times.

A second unit is currently being pressure-cycled down to 900-ft water depth in a seawater enclosure and is rated at over 100 amp.

A third design suitable for use as a multipole, self-aligning, remotely installed connector in deep water is being developed. It will be operable by guidelines or mechanical manipulator and a production version is hoped for trial this year.

Other devices coming. As a result of the trials and experiences other electrical devices were developed.

An oil-filled pressure-balance nickel-cadmium battery housed in a lightweight fiber-glass enclosure has been built and tested successfully in seawater.

Encouraging results have been achieved with a direct conversion thermoelectric generator assembly. This derives heat from crude oil passing through the wellhead pipework and cooling from seawater, thus producing dc power.

Nuclear Power

Submarines and submersibles. The dominant position of nuclear power for military submarines is now well established but as yet has not been an economic source for the small civilian marined submersibles. The only submersible built was the U.S. Navy, N.R.I., at a cost of nearly $100 million, which uses a steam turbine and nuclear raising plant.

Offshore and seabed power stations. With the urban economic and environmental demands on energy and power distribution, concepts have been prepared in the U.S. for offshore and subsea nuclear power stations.

This is not surprising when one considers that the knowledge is readily available. A nuclear submarine can be considered as a mobile power station. Fortunately for civilian application the mobility (perhaps only a buoyancy system for towing, installation, and recovery) complex navigation systems and of course missiles are *not* required and therefore such an application could be readily conceived.

6

Subsea Oil Production

FOR political, economic and logistical reasons man is searching farther and farther afield to assure his supply of oil and gas. Initially, this involved exploration and production in the shallow coastal waters; now exploration is moving further offshore and into deeper water.

The traditional approach of building man-made steel islands proves uneconomic in water depths exceeding 200-300 meters. Structure costs rise exponentially with increasing water depth.

The new approach locates production equipment on the seabed, making the facility less dependent on a fixed surface piercing structure. While the exact amount of equipment to be located at the mudline is currently subject to a number of restraints, two clearly independent philosophies are apparent with regard to system intervention. As with the exploration of outer space where manned and unmanned techniques are employed, so in the exploitation of the sea (inner space) using subsea equipment.

Where manned access is employed, the philosophy is based on the concept that one should not fight the subsea environment, but instead isolate man, and where necessary, equipment from it. Transportation between the surface and the seabed is by a service capsule or diving bell. Other developing systems rely heavily on remote intervention for installation and maintenance using robots or surface controlled handling tools and guide-line systems. Between these extremes systems are developing which provide for manned backup if required to supplement remote intervention techniques.

The evolution of the subsea oil system is likely to minimize, or perhaps eliminate the role of man subsea, but at the present time two facts remain.

1. The versatility of man is invaluable in dealing with the unforeseen, and his presence on the task is both time and cost saving, at least in the short term while remote intervention techniques are being perfected.

2. The development of remote intervention is underway and should bring production of oil and gas in depths beyond 200 meters, at a much earlier date than previously considered.

The Subsea Production System

A deep-sea production system of the future could comprise wellheads, flowlines, production and test manifolds, separators, pumps, compressors, dehydrating, demulsifying, injection and a power source. Artists' impressions abound on this subject and one more appears in Fig. 6-1, which is neither entirely factual nor fictitious, since although parts of the system are only conceptual at present, other parts are already in operation on live wells.

Today there exist several subsea production wells both on single and multiwell completions; knowledge on flowline laying and subsea connecting is improving, and designs are available for flowline manifolding. Installation, operation, and maintenance dealt with mainly by manned intervention, and the simpler day to day operations are remotely monitored and controlled from the nearest platform.

Currently the use of subsea completions in shallow waters allows the coverage of existing fixed platforms to be increased beyond the area that can be reached by deviated drilling. Also, single well completions can be employed to make use of exploration wells remote from a platform or as stepout wells for gas or water injection into the reservoir. Even multiwells have now been completed subsea with flowlines to the nearest platform. The downstream facilities shown in Fig. 6-1 are however still in the stages of concept and design and will take several years to mature.

In relatively shallow water, platforms and subsea production

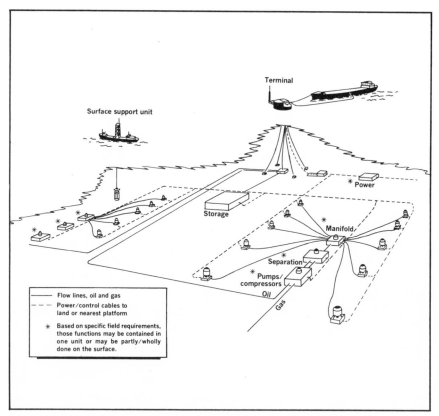

Fig. 6-1.   Impression of a complete subsea production system.

wells are often complementary and the placing of the entire production systems subsea is not generally warranted; but as production goes deeper and further from land the economic attraction of achieving a comprehensive and self contained subsea production system seems very attractive.

At present there is still reliance on the presence of a surface facility to provide functions such as well control, crude oil processing, gas flaring, accommodation for personnel, and perhaps tanker loading. The most vital role played by the surface facility is the provision of a power supply for control, oil pumping, and gas compression requirements.

105

Even the systems of the future will retain an interface with a shore terminal, tanker loading facility, and a power supply source. Also, some means of intervention for maintenance is required. This will probably involve a surface support vessel of a type that can also handle the installation phase.

While it must be recognized that an enormous amount of experience has been built up on offshore platform operations over the last 25 years, which is embodied in a high degree of specialization of men and equipment and the development of sound and safe practices, the placing of production equipment on the seabed would seem to have a number of economic, logistic and political advantages:

1. The considerable cost of fixed platforms in deeper waters is further increased, in "present value" terms by the protracted time it takes to design, fabricate and install the larger structures. The subsea alternative, by comparison, could be available on a relatively "off the shelf" basis.

2. The subsea oil system has far fewer foundation problems, provides no shipping hazard and does not require special on-shore construction facilities. (An advantage particularly in remote areas.)

3. The safety of a sea-floor system is superior since catastrophe should only affect one part of the system instead of the whole system, and storm damage would be very unlikely.

4. Optimum well spacing and drainage patterns can be achieved in shallow geological structures; in-fill wells, secondary recovery, or injection patterns are not affected by the nearest platform location. The need for deviated drilling can be avoided.

5. Subsea oil systems can be employed where ice floes exist and even where aesthetic considerations have to be recognized.

Maintenance of conventional land-based oil production equipment is well understood, and the specific maintenance requirements will not differ greatly for the subsea production system. However, as the access will be much more difficult and costly the whole approach to maintenance warrants reconsideration. An analysis of the frequency with which maintenance activities are repeated varies from 6 months to 10 years depending on the activity.

This and the frequency distribution provide the basis of maintenance planning on a preventive basis and the grouping of equipment with similar life spans in separate retrievable and replaceable nodules. Also double or triple redundancy can be employed to ensure that key components function satisfactorily for the full project life.

Operations, in the broad sense, subsea can be categorized into installation, connection, operation, and maintenance of equipment, and to achieve these objectives an interesting range of solutions has emerged. As with space exploration, the extent to which the versatility of man or the dispensability of machine is involved is a matter of degree, opinion, and compromise. In the subsea world, the intervention systems which have evolved so far for wellheads vary from one extreme to the other as shown on Fig. 6-2.

The Means of Intervention

The present status of evolution of the various means of intervention, shown in Fig. 6-2, is such that most are at least at the prototype trials stage, and several developments are already being employed on live wells. The advantages of each of these various developments depends on the actual operation to be carried out.

Subsea Well Equipment and Completions

The first known underwater completion (UWC) on the North American continent was made in 1943, in 35 ft of water in the Canadian waters of Lake Erie. Since then, more than 300 UWCs have been made in Lake Erie, and this represents the largest concentration of UWCs in the world. These have all been relatively low pressure (less than 2,000 psi) gas wells in shallow water (less than 85 ft). The wells are equipped with simple land type Xmas trees (Fig. 6-3) which require divers for installation, flowline connection and valve operation.

Lake Erie development is continuing. This program is an excellent example of the application of very economical completion equipment and methods to a specific set of known conditions. There

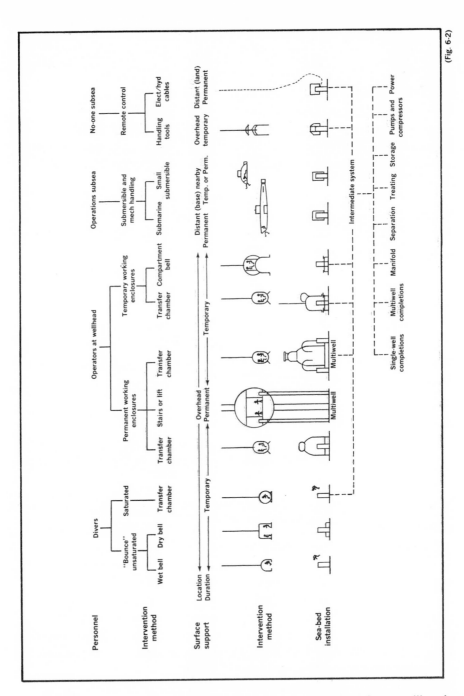

Fig. 6-2. Evolution of man and machine intervention on subsea wellheads.

108

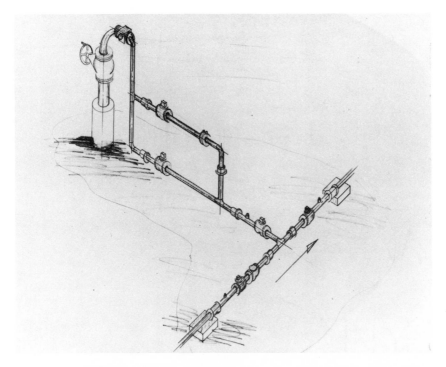

Fig. 6-3. Lake Erie wells are equipped with simple land-type Xmas trees.

has been no effort to develop the equipment and methods for water depths beyond diver capability, simply because the maximum water depth in Lake Erie is 210 ft.

Although UWCs were being made in Lake Erie as early as 1943, the development of deep water subsea wellhead equipment and completion technology was not seriously undertaken for the open sea until the early 1950s. During the mid-1950s, R&D work was initiated to develop remotely operated equipment and techniques for seafloor well completions. One of the early approaches, for application out to 200 ft of water, involved the use of a 10-ft diameter, one-atmosphere cylindrical chamber (Fig. 6-4) to enclose the wellhead. This method was discarded in favor of the wet tree approach.

This was followed by the development of Shell's remote universal drilling and completion system (RUDAC) using a guideline sys-

Fig. 6-4. One-atmosphere cylindrical chamber encloses Lake Erie well-heads.

Fig. 6-5.   *First known underwater completion in the open sea, December, 1960, off Louisiana*

Fig. 6-6.   Shell's manipulator-operated system and its associated MOBOT.

tem for maintaining contact with the wellhead during drilling and completion. This resulted in the installation of the first known (December 1960) UWC in the open sea (Fig. 6-5) on the outer continental shelf off Louisiana without reliance on divers.

This well was the first to have all equipment operated from the drilling rig during the completion operations. Concurrent with the development of the RUDAC system was the development of Shell's manipulator-operated system (MO) and its associated MOBOT (Fig. 6-6). This system did not require guidelines and was used in the Santa Barbara Channel (Molino field) during the period 1963 to 1965.

Between 1960 and 1974, some 106 subsea wells have been completed on the offshore continental shelves of the free world. These completions are distributed throughout the world, with 36 in the Santa Barbara Channel, 46 in the Gulf of Mexico, 9 in the Persian Gulf, 7 in the North Sea, 4 in the South China Sea, 2 in the

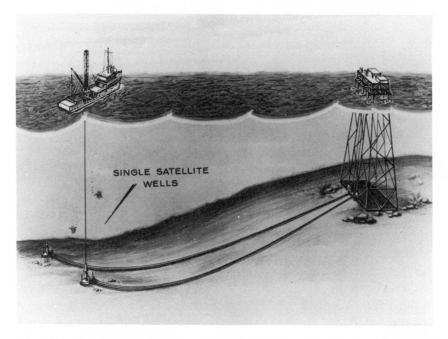

Fig. 6-7. Nearly all underwater completions have fail-safe valves operated from nearby surface (shore or platform) installation.

113

Mediterranean off Spain and one each in the Arabian Gulf and off West Africa (Gabon). Of these, seven were experimental for equipment evaluation and 99 were completed as either producers or water injection wells.

The completions were made by some 16 companies or groups of companies in water depths ranging from 50 ft to 375 ft. The equipment used on these wells has required a varying degree of reliance on divers. Nearly all have fail-safe valves operated remotely from a nearby surface installation on platform or shore (Fig. 6-7) using hydraulic control systems. Although a few have been equipped for remote connection of flowlines, all but one have required the services of divers to assist in this function. To date, no installations have been made in water depths beyond diver capability.

The Zakum Subsea Production Scheme

The project known as the Zakum Subsea Production Scheme covers the period August, 1969 to April, 1972, in which time equipment and techniques developed by BP, CFP, oilfield manufacturers, and service companies were operated and evaluated.

The objectives of the scheme were:
1. To achieve production from the well at Zakum 39 using subsea methods.
2. To gain experience of subsea production methods and equipment and thereby evaluate their application to future deepwater operations.

The range of equipment and techniques investigated covered most aspects of subsea production. In addition to primary production equipment (wellhead, valves, actuators, flowlines), a range of ancillary equipment and support services was also included (separators, underwater power sources, instrumentation, diving services, surface support vessels).

The operating conditions at Zakum were not demanding, as the water depth was only 22 m, and this allowed easy diver access. A contract diving service provided the support to the subsea production group, which comprised production, petroleum, and instrumentation engineers who had been trained to dive.

Equipment—Design Details and Performance

The well and wellhead assembly. Well ZK-39 was drilled as a development well in the Zakum oil field and completed with a subsea wellhead in order to provide facilities for testing the various items of production equipment. Throughout the evaluation period, the well was treated as a normal field well and, with a few exceptions, was always on production or available for production. Total well production, to the end of the scheme, was 1.7 million barrels.

Little trouble occurred with the well or the wellhead assembly during operation, and the problems that did occur would be expected in normal oilfield operations and not specifically due to operating in a subsea mode.

However, when problems did occur, it took appreciably longer to rectify them. This was partly due to the difficulties of access and partly due to the development nature of the scheme, in that maintenance routines had to be established.

Proper routines for maintenance and repair of subsea wellheads, and indeed all subsea equipment, are regarded as a priority to the success of any future subsea systems.

The oil from the well was piped through a manifolding system and finally discharged into the flowline leading the ZK.6 junction well. The manifolding allowed the diversion of oil through any item of test equipment (e.g., separators). The manifolding was constructed of 4-in.-diameter rigid pipe sections.

Connections in the manifolding system were mainly by 4-in. clamp-joints, but although they gave satisfactory service, there are reservations about their use on future subsea work. A particular disadvantage of the joints is the difficulty of attaching cathodic protection anodes.

Connections between the manifolding and the test equipment, and also the well, were made by flexible hoses. These were very successful, although a certain amount of care has to be exercised in their handling. The use of flexible hoses reduced the requirement for the precise positioning of the test equipment (some of which weighed 60 to 80 tons) and allowed easier assembly of the clamp-joints.

The scheme included the design and evaluation of techniques for installing, recovering, operating, and servicing subsea separation systems. Briefly, investigation was made into: (1) installation and recovery systems, (2) separator control systems, and (3) means of servicing the control systems.

Two over-all design concepts were implemented for inclusion in the scheme. System A, Fig. 6-8, comprised a two-component unit. A heavy anchor section gave stability to the whole unit on the sea-bed and also provided permanent oil and gas flow connections to the manifolding. A separate process section carried a separator and a bell which contained pneumatic level control and radiation-detection level-control systems, control valves, and an electro-pneumatic safety shut-in system.

Installation of System A consisted of towing the complete unit to site, separating the two sections, and lowering the anchor section to the seabed under the control of winches mounted on the process section. Then, after flooding the anchor section, the process section was pulled down using the same winches, and the two sections mated together again. Recovery, either of the entire unit or just the process section, was the reverse of this procedure.

Servicing of the control systems was generally carried out by divers working "in the wet," though for special work the bell could be "blown" with air or inert gas to provide a dry atmosphere. Clearly, this system was totally dependent on divers and, as such, would be limited to installation at depths of say 100 m at which the divers could operate satisfactorily and economically.

System B, Fig. 6-9, employed a single unit construction, and in order to obtain stability in all phases (towing to site, installation, and operation), two side stability floats were mounted on movable arms attached to the separator vessel. A conventional pneumatic control system, with the gas supply and all of the control valves, were mounted in a removable capsule that could be recovered for servicing.

Installation was simply a matter of towing the unit to site and adjusting the buoyancy and the position of the side stability floats as required for each phase of the operations. Although divers played a large part during the installation, inspection, operation

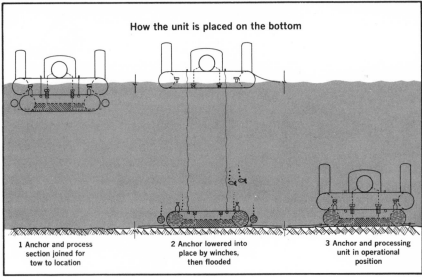

How the unit is placed on the bottom

1 Anchor and process section joined for tow to location

2 Anchor lowered into place by winches, then flooded

3 Anchor and processing unit in operational position

Fig. 6-8. First subsea separator, built by National Tank Co., is in operation in Arabian Gulf's Zakum field.

117

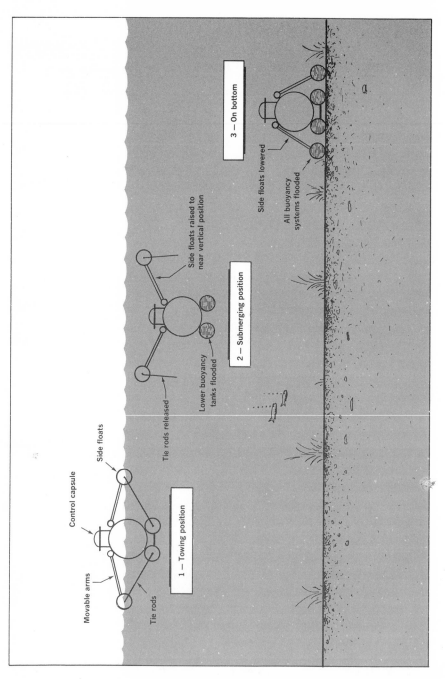

Fig. 6-9. Installation sequence for System B.

and servicing phases, it was hoped that this design concept could provide the first step toward a diverless system suitable for deep water.

Figs. 6-10 and 6-11 show some of the Zakum operations.

The main conclusions obtained in this phase of the scheme suggest that it will be some time before subsea separation systems are used. The development will be gradual, in that they will generally be associated with an offshore structure or a subsea complex and not required to operate in isolation. In a subsea production system thought would have to be given to the location and operating pressures of the separators. Installation of the main units would not appear to present any problems, but the main point seems to be whether it is necessary to provide individual flotation systems permanently attached to each unit.

In order to obtain reliable subsea separation, there must be further development to achieve satisfactory control systems, since the pneumatic system was unreliable and unsuited to deep-water operation. An electric or electro-hydraulic system is preferred, but simple mechanical systems could be investigated due to their inherent reliability. The successful operation of a subsea separation system will also involve a wide range of other requirements which must be satisfied (automatic flowline connectors, reliable electric connectors, servicing methods, etc.).

In order to dispose of gas produced from the separators on the test site, two disposal systems were evaluated.

A flare buoy was successful, although the instability of the buoy was a problem. Both principle and construction are simple, and the unit would be adaptable for deepwater operation.

The dispersal system, which diffused the gas in seawater, is simple and reliable, but it would present ecological and safety problems during operation as the gas is absorbed in solution and not burned.

The requirements of the installation included the supply of electricity for the valve actuators at the wellhead, for a back-up power supply to the oil/gas separators, and for wellhead instrumentation.

Various forms of power source were evaluated and these all supplied power to a large nickel cadmium battery of 300 amp-hr effec-

Fig. 6-10. *Diver working on subsea wellhead in Zakum field.*

Fig. 6-11. Subsea separator for the Zakum field in Das Island harbor.

tive capacity, housed in an enclosure sealed at atmospheric pressure. A voltage-sensitive relay was included in each power input connection to disconnect the charging source as the battery terminal approached the electrolyte gassing voltage of 1.37 v/cell. Subsequent chemical tests showed that with this system virtually no gas generation occurred, and hence there was no electrolyte loss over duty periods in excess of 6 months.

The power sources evaluated and/or installed, included the following:

- A 20-W radioactive isotopic thermoelectric generator. This functioned very well for the complete duration of the scheme.
- An oil expansion, glandless, turbo generator of 600-w capacity operating at oil system pressure. This was not installed on site in a submarine form, since considerable experience had already been obtained.
- A gas expansion turbo generator of 4-KVA capacity designed

121

as a glandless unit to operate on gas stripped from the crude oil by the separators. This unit was built in submarine form and extensively tested on land but as yet has not been installed under water.

- A small section of a thermoelectric generator deriving power from the temperature difference between the geothermal heat of the crude oil and the surrounding sea water. This unit, although too small to supply power to the battery, was installed for environmental evaluation purposes.
- A vortex generator, using the Ranke Hilsch effect to derive a temperature difference in gas, and thermoelectric lead tellurium nodules, was also built and run as part of the oil/gas separator package in System A.
- A multicore submarine cable 2½ miles long was laid from an existing wellhead platform carrying an adjustable dc charging supply derived from an existing oil expansion turbo generator. This cable also included 40 conductors for control and instrumentation.

Power-consuming devices included wellhead temperature and pressure indicating transducers with conventional components housed in atmospheric pressure enclosures.

A prototype valve actuator was designed and installed to operate in 4-in. gate valve as a wellhead isolating valve. This unit was based on a dc stepping motor, brushless, operating in an oil-filled enclosure at ambient pressure. The problems of close travel limits switching tolerances and easy installation by divers were adequately overcome. The many improvements that became apparent would be included in subsequent models (increased speed, lighter construction, better marine fouling protection, etc.).

However, after 18 months' submersion the prototype was in excellent condition. It had, on occasion, operated the wellhead valves satisfactorily by remote control via a radio link and submarine cable from a control room 55 miles away.

Interconnections between submarine equipment were made by trailing, pliable-armored polychloroprene sheathed cables and by using a proprietary plug and socket connector. It was the repeated failure of the connectors after a few weeks in service that caused the complete disruption of the electrical installation.

Apart from the connector, all components installed in the scheme worked and were proved to be satisfactory in principle whether oil was immersed and balanced to ambient pressure, or housed in atmospheric pressure enclosures. In view of the unfortunate experience with the connectors, and there being no others available, BP has designed, developed, and patented an oil-filled submarine plug and socket that has been used successfully in underwater trials over long periods and tested at a great depth.

An investigation is at present being pursued into the operation of batteries within oil-filled enclosures balanced to ambient pressure.

Although thought was given at the outset to the elimination of electricity from the Zakum trials, experience has shown that electricity will usually be required to some degree on submarine installations.

Communications and instrumentation. In order to monitor operations at the test site a communication and instrumentation system was incorporated. The system was simple and operated on a conventional dc electrical basis. Indications of wellhead valve actuation, alarm systems for the separators, pressure, and temperature readings were transmitted to the nearby wellhead structure via circuits in the multicore submarine cable, and then via the conventional microwave link to a display in the land-based control room. Limited remote control of the facilities at site was provided by the same transmission system.

Acoustical and inductive systems were developed and evaluated off the site, but as they were not reliable, they were not installed. Development is continuing on both aspects and if successful could have significant advantages for underwater communication and control compared with submarine cable systems.

Wireline work. The first series of wireline operations was carried out during the first 3 months of 1970. The support vessel used was the M. V. "AJAX," a 165-ft-long, forward-bridge workboat suitable modified.

Three methods of performing wireline work in a subsea well were tried.

1. Using a conventional lubricator mounted on the subsea well-head and a compensating system mounted on the deck in order to remove the effect of vessel movement on wireline operations. This simple method was very dependent on divers.
2. Using a flexible riser (in effect a long lubricator), which "extends" the wellhead to the surface. The flexible riser must be maintained under tension in order to keep it reasonably straight; this requires a compensating system, which is also required for the actual wireline operations.
3. Using a completely encapsulated submersible winch system, guided to and placed on the wellhead, and operated by remote control.

The main objectives of the trials were to assess the practicability of each method, to compare the methods (in an effort to assess the likely areas of application and economics), and to provide guidance for the further development of the equipment.

Following on from the 1970 wireline operations, a small swell-compensating unit was developed and constructed as part of the continued development of the conventional lubricator system.

The unit was completed in mid-1971, and initial tests were carried out in the Mediterranean Sea. Following shipment to the Arabian Gulf, wireline operations were carried out at ZK.39 in early 1972.

From the results of the 1972 wireline operations, it was concluded that it was possible to carry out reliable wireline operations from a floating vessel using the conventional lubricator method.

The dependence of divers, however, at the present time, limits this method to operations in a maximum water depth of 100 m.

It is considered necessary to develop the subsea aspect of the equipment in order to reduce the requirement for divers. If this can be done (the development should be a simple matter), then it is anticipated that this system could be used down to 150 m water depth. At greater depths than this, there may be two problems:

(1) diving operations become very expensive, and (2) simple guideline systems may fail.

With development of a more sophisticated wellhead connector, the system appears adaptable to greater water depths provided that (1) guidelines are used, and (2) the effect of current drag on a long exposed section of wireline does not significantly influence the wireline operations.

Ultimately, the conventional lubricator system will have a depth limitation, and it will be necessary to consider alternative methods of doing wireline work. Such methods include: (1) the flexible riser, (2) the submersible winch, and (3) the atmospheric systems (manned capsule approach).

The ability of the surface support vessel to maintain position is of the utmost importance to perform wireline operations successfully by any of the methods described above.

For deepwater operation, the vessel would need to be large and be fitted with a range of equipment for servicing a subsea oil field. The method of development of the field and general servicing system will govern, to a large extent, the choice of the wireline system.

Current subsea Xmas tree types. Xmas trees are installed at the wellhead to provide well shut in and flow control functions. Subsea trees can be divided into two basic types, the "wet" and the "dry." The "wet" type (Fig. 6-12) has all components exposed to the sea and must be serviced by specially designed manipulators, conventional divers, or by recovering the tree or components to the surface. The "dry" type has all tree components housed in a chamber (Fig. 6-13).

This system can be serviced by men working in an atmospheric pressure air environment on the seafloor. To date only one chamber system has been installed on a live subsea well.

Typically, the wet tree is completely assembled and tested before installation from a drilling rig. The tree is run and latched to the seafloor casing housing with a hydraulic connector controlled from the rig floor. After the rig has moved away from the well, flow-lines are connected to the Xmas tree, either by divers or manipulator techniques controlled from a surface vessel.

Fig. 6-12. "Wet" type of subsea tree has all components exposed to the sea.

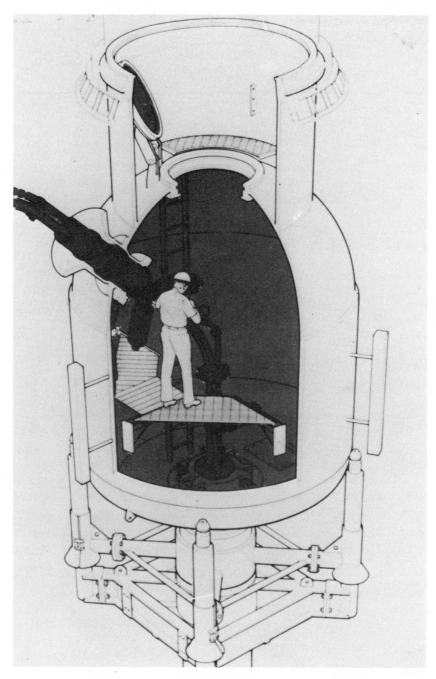

Fig. 6-13. "Dry" type of subsea tree has all components housed in a chamber.

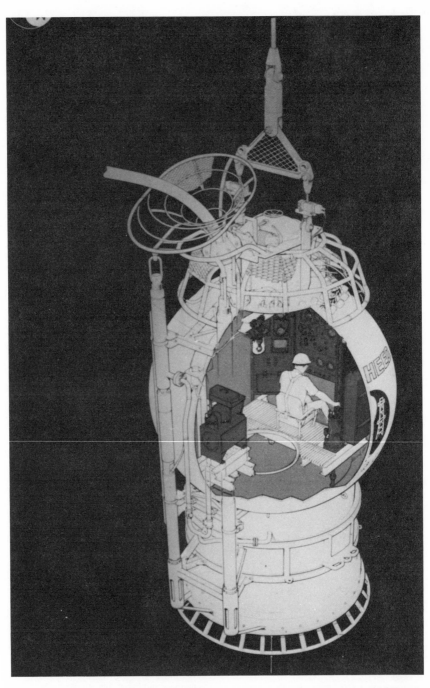

Fig. 6-14.   Downhaul utility capsule is used to transport men to and from the chamber at atmospheric pressure.

With the dry chamber system, the chamber is run from a drilling rig with the tree components disassembled and stowed inside. A hydraulic connector, similar to that used with the wet tree, is used to attach the unit to the seafloor casing housing. After the rig moves off location, a service boat is moored near the well and the downhaul utility capsule (DUC), Fig. 6-14, is used to transport men to and from the chamber at atmospheric pressure. The men, using a cable and puller assembly, effect the flowline connection prior to the assembly and testing of the Xmas tree.

Control systems for both types of Xmas trees can vary from a simple hydraulic system wherein all fail-safe valves are operated at the same time to more sophisticated electro-hydraulic systems. The simplest hydraulic control systems provide remote actuation of the subsea valves. The valves are closed automatically by spring force (fail safe) when abnormal pressures occur in the flowlines or the hydraulic control system, or when malfunctions of the surface equipment occur. The more sophisticated electro-hydraulic control systems use multiplexed control systems to provide individual valve control, valve position indication and subsea wellhead pressures.

Present State of Subsea Completion Capability

For a subsea well, completion capability can be divided generally into five categories which are:

- Installation of tree—wet trees can be installed in water depths commensurate with floating drilling rig capability. Wells have been drilled, but not completed, in water depths to 1,500 ft and one well is currently being drilled in 2,150 ft of water. The chamber system is currently limited to 1,200 ft because of the depth rating of the DUC.
- Connection of flowlines to tree—a number of remote flowline connection systems for wet trees have been built and tested. All but one have utilized divers to assist or to observe. For the lone chamber installation, divers were only required to make visual observations and this function will be performed in the future by underwater TV.
- Tree maintenance—wet tree maintenance operations have been successfully conducted using divers. Repairs requiring recovery of the tree have been made using a drilling rig. Tree repairs in

the chamber system have been successfully accomplished using Lockheed's DUC atmospheric diving system. There should be no significant problems in servicing trees in chambers out to the water depth limit of the DUC.

- In-casing maintenance—work involving the removal of the production tubing strings will require drilling rigs. Flowlines must be remotely disconnected and wet trees must be removed in order to pull the tubing. Production tubing pulling is effected by chamber removal or by pulling through the chamber after the disassembly in flow systems with dry trees.

- In-tubing maintenance—conventional wireline and through flowline (TFL) hydraulic pumpdown methods have been used to perform in-tubing maintenance in subsea wells. The use of wireline methods from a surface vessel through wet trees is limited to relatively calm sea conditions, but this method has been used in a water depth of 240 ft in the Santa Barbara Channel.

Wireline methods can also be used inside a chamber providing provisions are made for venting of any gas or oil leakage. While wireline servicing has been the traditional tool since the last war, TFL development has paralleled subsea completion development in recent years. The TFL method offers the most potential for in-tubing maintenance of deepwater subsea completions (both wet and dry chamber). It is simpler and more reliable because of the absence of any heave compensation equipment and less chance of leakage, and should also be lower cost. Fig. 6-15 shows a typical tool string and Fig. 6-16 a circulation system.

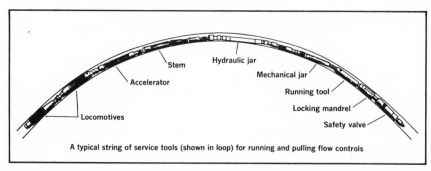

A typical string of service tools (shown in loop) for running and pulling flow controls

Fig. 6-15. Typical tool string used in TFL maintenance work.

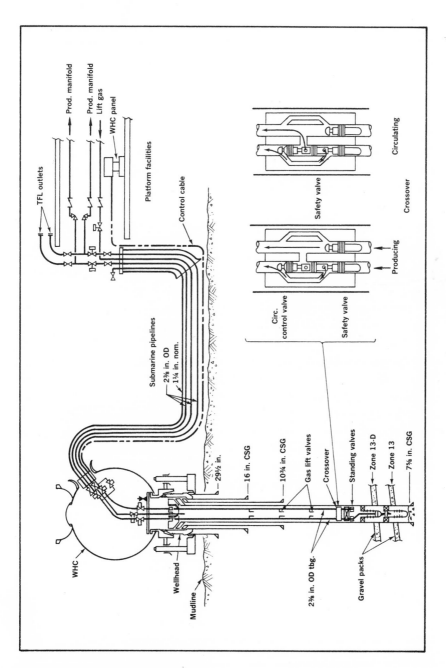

Fig. 6-16. TFL circulation system.

Considerable TFL experience has been gained in subsea completions in the Gulf of Mexico. This method has been used to pull and rerun various subsurface flow controls, including safety valves, circulating control valves, standing valves, gas lift valves; to shift sliding sleeves; and to successfully perform an acid job in one zone. Development work is under way to provide through tubing remedial techniques, including perforating, sand consolidation, sand washing, formation squeezing, and cement plug backs. Most TFL experience has been in 2-in. tubing, but systems are being developed for 3-in. and 4-in. and larger sizes.

Prototype Subsea Production Systems

A number of production systems for subsea wells are in various stages of development. Four prototype systems planned for installation have actually reached the fabrication stage. These are being developed by Deep Oil Technology, Subsea Equipment Associates Ltd. (SEAL), Exxon, and Lockheed (for Shell).

The Deep Oil Technology subsea system consists of a cluster of subsea wells and a production facility mounted on a common template located on the ocean floor. The equipment, which is designed for 1,600 ft of water, will be serviced with a manned submersible diving bell equipped with manipulator arms designed to make and break connections during maintenance operations.

A multiplex control system is used for the remote operation of equipment during normal production conditions. Production from the subsea facilities could be directed to a platform, to shore or to a floating facility such as a tension leg platform. Four commercial wells were completed in the Persian Gulf in 1972 using portions of this system.

The Subsea Equipment Associates Limited, (SEAL), approach consists primarily of two different subsea systems (see Fig. 6-17):

1. The multiple well manifold production station (MWMPS), and
2. The single wellhead completion (SWC).

These systems were designed to accommodate varying water depths (1,500 ft for prototype) and field sizes, most economically.

Fig. 6-17. SEAL system for underwater production.

SEAL employs a combination of "wet" and "dry" concepts in their two systems as the SWC is a wet tree application and the MWMPS is primarily a dry system. The subsea working enclosure (SWE) of the MWMPS (Fig. 6-18) has a dry one atmosphere inert gas environment for safety considerations.

This SWE houses well controls, valve manifolding monitoring devices, through-the-flowline (TFL) equipment, test separator and pigging equipment. The SWC features a "split" Xmas tree for economic maintenance and is a combination of the manned and unmanned approaches for subsea intervention. The concept employs a replaceable nodule with a remotely operated handling/service tool for normal operations. For rare operations (once or twice in the life of well), and as a back-up facility, a manned intervention at atmospheric pressure is possible.

Equipment in both systems is controlled subsea by remote operation via power and telecommunications link and maintenance is achieved using trained oilfield personnel.

Field testing of the MWMPS was initiated in 1972 in the Gulf of Mexico in 250 ft of water. The SWC has been rigorously tested at company sites in the Mediterranean comprising shallow and deep water testing. A version of the single wellhead completion, the first commercial application, was ordered for installation in the North Sea during 1975. The initial phase (see Fig. 6-19) was accomplished in May 1975 to be followed subsequently by the installation of the production control assembly and the laying of the flowline.

The flowline will be connected to a nearby platform, enabling one of the salient features of the SWC; i.e., the early production of hydrocarbons, to be realized with the attendant economic payoff.

Exxon's subsea production system (SPS), Fig. 6-20, concept consists of a cluster of subsea wells and associated production controlling, separating, and pumping equipment mounted on a subsea template structure which is designed for application in water depths out to 2,000 ft. The production fluids are transported to surface via pipeline to shore, to a platform or floating facility. The subsea equipment is remotely controlled using an electro-hydraulic supervisory control system.

Pumpdown (TFL) tools are used to service the wellbore equip-

134

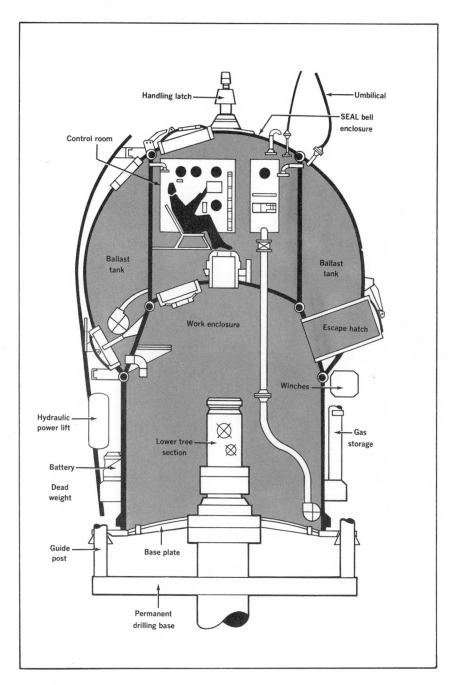

Fig. 6-18. Subsea working enclosure for SEAL system.

Fig. 6-19.   Single wellhead completion for the North Sea.

ment and a mechanical manipulator (Fig. 6-21) operated from the surface is used for maintenance operations on the subsea structure. Equipment on the structure is especially configured in nodules so that system replacement can be achieved using the manipulator. The test installation of this system was made in the Gulf of Mexico in November 1974.

Lockheed Petroleum Services Ltd. has now completed the second phase in its development of a subsea production system. The

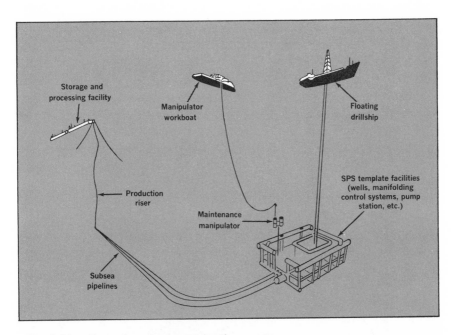

Fig. 6-20. Exxon's subsea production system.

Fig. 6-21. Mechanical manipulator for TFL pumpdown tools.

137

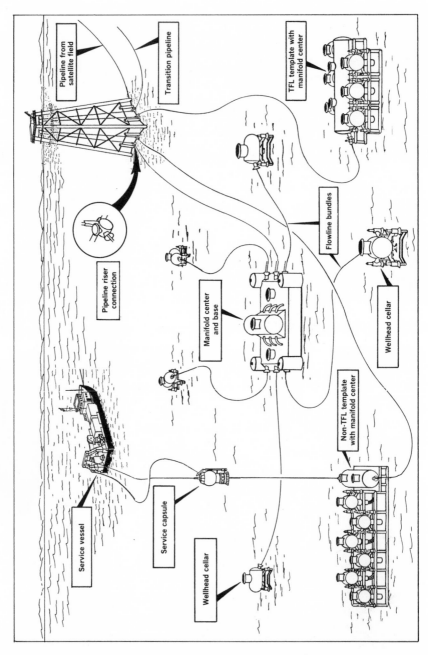

Fig. 6-22. Lockheed subsea production system.

first phase was the installation of a single encapsulated wellhead for Shell Oil Company in 1972, described earlier. The second phase was the installation of a manifold center capable of combining the production from three satellite subsea wellhead cellars. This manifold center completed its land tests during the summer of 1975 and was later submerged for Shell in the Gulf of Mexico's Eugene Island area.

Lockheed's ongoing development plans, pictured in Fig. 6-22, include templates for directionally drilled wells capable of production with or without pumpdown (TFL) tools and a pipeline riser connector chamber. The template approach permits drilling of several wells from a single anchoring of the drilling vessel and reduces the number of flowlines needed on the ocean floor by commingling production in a built-in manifold.

The pipeline riser connector chamber makes it possible to weld ocean-floor pipelines to a platform's riser pipe using dry-land welding techniques. All operations with the Lockheed system are carried out in a dry, one-atmosphere, shirtsleeve environment. All components in the Lockheed system can be remotely controlled by hydraulics, electronics, or a combination of these.

7

Communication and Navigation

FOR the diver or submersible pilot to explore and work upon the seabed, the two functions of communication and navigation are important to ensure the safe, effective and economic use of the time spent on the task. Sound underwater plays a vital role in a wide range of activities from hydrographic and geological survey work to submarine and fish detection.

Of particular significance offshore has been the development of dynamic-positioning systems for drill-ships. The capability to drill in depths of waters of 3000 feet or more now exists.

Diver communications. Hand signals (Fig. 7-1) have been used by divers for many, many years and no doubt will remain a basic part of his communication system. Similarly line-pull signals are simple and effective for exchanges between the tender on the surface and the diver below.

These systems are, however, limited for hand signals by visibility (a diver's vision even under favourable conditions seldom extends beyond 100 ft and in coastal inland waters it may well fall to zero or a few feet) and for the line-pull system by depth.

Therefore, other means are essential to communicate large quantities of data in real time for expediting tasks and ensuring safety of divers, vehicles, support vessels, and bases ashore.

For divers operating on surface demand communication can be made by surface to diver telephone.

For a free-swimming diver (no umbilical) the best primary communication method is by the acoustic transmission of the voice through the water, rather than by transmission along a wire con-

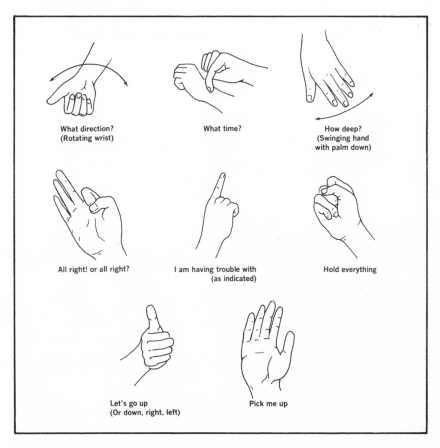

Fig. 7-1. Diver hand signals used by the U.S. Navy.

ductor telephone. The acoustic transmission system has the advantage of allowing free-swimming diver-to-diver and diver-to-surface ship or submarine voice communication without umbilicals.

Typically, this equipment has operating ranges of 500 meters at depths of up to 90 meters and operates on the standard underwater telephone frequencies. One such item of equipment is thigh-mounted and operates for two hours without requiring battery renewal. A bone conductor transducer is used instead of earpiece and microphones, enabling the diver to wear a normal helmet. See Fig. 7-2.

In addition, for location of free-swimming divers, a small acous-

Fig. 7-2. Diver equipped with a 40 kHz through-water telephone.

tic transponder is available for clipping on his belt. With the use of a normal acoustic location system his progress may be simply monitored.

The process of forming words and speaking clearly is more difficult under water for several reasons. There is the problem of a limited volume of air in the face mask, and therefore when completely submerged there is a negligible transmission through the walls. To overcome this problem oral/nasal covers are introduced or an amplifier and loudspeaker.

With dives to greater depths employing the use of diving bells and helium breathing mixtures, a further problem on voice communication for the diver is encountered. Due to the different density of the mixture in substituting helium (compared with the nitrogen in compressed air) human speech is distorted causing the "Donald Duck" or helium speech effect.

Signal processing units are available either as separate units or as part of the underwater telephone equipment for unscrambling the helium speech. One unit, the ARL (Admiralty Research Laboratory) helium speech processor is of particular interest since it may readily be installed in submersibles with diver lock-out facilities.

Submarine detection. The French Professor, Langevin, first developed active detection by ultrasonic sound pulses in 1916. The term "sonar" (sound navigation and range) is of American origin, but the science was further developed by the Royal Navy during World War I under the Anti-Submarine Detection Investigating Committee which gave rise to the acronym ASDIC. Since then, there has been a long history of continued development.

However, as the techniques of sonar have improved, so also, and in greater measure, has the performance of the submarine, and today its detection still presents one of the greatest problems in undersea warfare. The speed of the nuclear-powered submarine, and more especially the long range of the tactical missiles with which it can now be armed, pose a need to detect it at longer ranges and greater depths.

To a limited extent, performance can be improved by increasing the power and lowering the frequency of the sonar but a major limitation, as is well known, lies in the refraction of underwater sound in sea water, which is very sensitive to temperature gradients. In many cases, a layer of warmer water near the surface give rise to a "shadow zone" beneath it and this results in a sharp limitation of the range of detection.

To overcome this, studies are being carried out to exploit the refraction, rather than its being an embarrassment. For example, sound can be deliberately bounced from the seabed on its way to and from the target, although to compensate for the energy losses involved requires large and expensive installations. To overcome this sonars could be placed at great depths, so that there is a good propagation path between them and the submarine (the "reliable acoustic path") but this again presents severe engineering problems.

These systems are perhaps unlikely to have much direct application outside defence. Nevertheless, their assessment and exploitation requires a deep understanding of how such modes of propagation are influenced by the physical properties of the ocean —the temperature and density distribution, the depth and nature

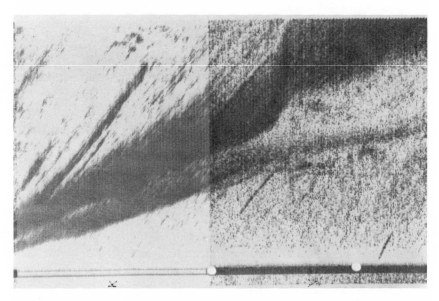

Fig. 7-3. Echogram from Simrad SU Sonar showing location of fish shoals.

144

of the seabed, and so on. Moreover, it is surprising what other interesting information can be revealed by high power sonar.

Besides this concern with detection at long range, there is the need to be able to detect and recognize objects at closer quarters. The techniques for this are naturally capable of being used to detect other objects, such as fishing trawls, and indeed shoals of fish themselves, (Fig. 7-3), wrecks, underwater installations, and the topography of the seabed.

Navigational Aids

Navigational aids are required for a number of purposes, e.g., positioning fixed and floating platforms, positioning of ships performing survey work, pipeline location, etc. There are several types of navigation systems for short ranges, up to 20 km, to long range distances over 2500 km.

Inertial navigation. Systems of this type may be readily fitted to both ships and submersibles. They have the capability of giving an accurate position relative to a known starting point, typically plus or minus 5 meters over a period of 3 months or more. The main disadvantages of inertial navigation systems are their high initial cost and the fact that once switched on at the starting point, the system must be left running continuously.

The costs are steadily reducing and one of the most competitive equipments manufactured by Ferranti is now being installed on merchant vessels for a cost of about double that for a satellite navigation system. It is claimed that short-term accuracies of plus or minus 1 meter may be achieved with this particular equipment and this should have application in the manned submersible market.

Radio navigation systems. These may be conveniently divided into three categories: hyperbolic radio systems, direct ranging radio systems, and orbiting earth satellite radio systems.

Hyperbolic radio systems. These are often known under the names of Sea Fix, Hi Fix, Loran A or C and Omega. The principle of operation consists of shore-based Master and Slave transmitting stations. The time difference or phase difference between the signals from two stations are compared and provided a line of position on a map. A similar comparison is then made for the signal arriving from a second pair of stations and a second line of position plotted.

The intersection of these lines fixes the position of the receiver. The short-range equipment has a typical operating range of 300 km over the sea during daylight hours with an accuracy of plus or minus 4 meters. The longer range Loran-C equipment allows position fixing over distances up to 2500 km with an accuracy of plus or minus 60 meters.

All of the systems require the accurate positioning and setting up of a number of transmitting stations on land as baseline reference points. Because of the changes to the earth's atmosphere during the nighttime, these systems are sometimes subjected to interference from signals reflected from the ionosphere rendering them unreliable at night.

Direct-ranging radio systems. These systems operate on the radar principle, i.e., time taken for a pulse to travel between a fixed station and the point requiring positioning. They normally operate at microwave frequencies and are limited to line or sight paths, 50 km or so, with an accuracy of plus or minus 1 meter. By the use of an airplane, to increase the line of sight path, ranges of up to 200 km may be covered. A typical use for this system on an offshore platform is as a docking radar for a loading buoy.

In each case, a transmitting base station requires to be set up.

Orbiting earth satellite radio system. Position fixing by this method requires the use of a radio receiver and computer only, the position being computed from the doppler frequency shift of a signal generated by the satellite as it moves overhead and the positional information transmitted by the satellite every two minutes.

For a single satellite pass, the position may be determined to an accuracy of plus or minus 20 meters and following several satellite passes, plus or minus 20 meters. One difference with this type of system is that the update in positional information is not continuous, unlike those of previous systems. However, for general marine usage the flow of information could be considered as continuous.

In principle, the use of pulses of acoustic signals through the water in sonar is analagous to pulses of electromagnetic waves through the air in radar.

146

Unlike radio navigation systems, however, sonar navigation is affected by the weather, i.e., sea state—the worse the sea, the less the navigation range. This is true for both surface and submersible vessels.

Acoustic transponder navigation systems. These systems consist of a number of acoustic transponders located in known positions on the seabed. An interrogation pulse transmitted from the vessel causes each of the transponders to "reply" to the vessel. The received signals are fed into the computer which determines the vessel's position relative to the transponders from the time taken for each signal to arrive at the ship.

The computer can provide a continuous plot of position for short range systems (a range up to 20 km) with a positional accuracy of plus or minus 1 meter. To this, must be added the inaccuracy in determination of the transponder's absolute position.

Transponders for this use have a life of 3½ years before requiring battery replacement. Most equipment of this type is too large for use on submersibles, but one system currently in production for the Royal Navy is much smaller and is likely to be of use.

Acoustic pinger systems. To aid relocation of a wellhead or pipeline, etc., an acoustic transponder is often moored on the bottom. It may also be associated with a pop-up buoy. Operation of a relocation receiver or suitable sonar causes the transponder to transmit a pulse, and hence the vessel may be located over the transponder.

Also available are pingers which transmit acoustic pulses at regular intervals, marking the position. Pingers have the disadvantage of a short battery life.

Neither system gives a very accurate positional fix since in water depths of 200 meters, a slant angle error of 5° will produce a positional error 1f plus or minus 16 meters.

Platform positioning system. Although not originally designed for this purpose, a Royal Navy positioning system, now released for civilian use, has the capability of positioning a platform within plus or minus 0.3 meters of any required point on the bottom. It requires the location of a single acoustic transponder on the centre point (or known offset point) together with four hydrophones fitted on the legs or base of the platform. Readout of platform position is continuous and blackout due to the acoustic noise caused by the manoeuvering of the surface vessels would not occur.

Long range side-scan sonar (project Gloria). In 1958 the National Institute of Oceanography (now the Institute of Oceanographic Sciences) pioneered the use of side-scan sonar to study the sea floor. Since then it has proved to be a powerful tool for delineating the patterns of sediments and rock outcrops on the floor of the Continental Shelf. The technique is used in various forms by a number of different research institutes and Universities, and can be employed for hydrographic studies of sand waves and by oil companies for survey work and pipe laying.

As a result of this success the N.I.O. decided in 1964 to undertake a study of the possibility of using a similar technique in the deep ocean to investigate geological features. Because of the depths of water involved, about 30 times that of the Continental Shelf, slant ranges of the order of 22 kilometers (12 miles) were contemplated in order to keep the same geometry. This meant high acoustic powers with low frequencies to reduce absorption, plus a hydrodynamically stable vehicle in yaw which would remain steady over the long 30 second, pulse repetition period needed to reach and return from 22 kilometers.

In 1969 the long range system was completed and went on trials at sea. The array of transducers, working at 6.5 kHz, is housed in a rotatable beam inside a glass fibre stream-lined vehicle which can be towed at a depth of about 300 feet. This enables high powers of up to 50 kw to be dissipated at depth without cavitation, it helps to decouple the array from the yaw of the surface towing ship and it places the array in a relatively quiet environment for listening. The vehicle is fitted with a number of controls and services. It can be remotely flooded and deballasted from its own high pressure air system and it has a yaw gyro which provides a reference for a servo loop which 'locks' the receiving beam to the bearing of the last transmission. The array of transducers can be remotely rotated so that the sound is launched and received at the correct despression angle for the prevailing propagation conditions.

The sonar has been used in 1969, 1970 and 1971 to reconnoitre

Fig. 7-4. The GLORIA towed sonar vehicle in its davit on RRS 'Discovery'.

and study certain geological areas and features in the Mediterranean and Atlantic. In 1971 an analogue correlator was used for
the first time giving improved picture quality and range. In the same
year the first saturation survey was attempted, mapping by over-
lapping records a 25000 sq km (7000 sq mile) area near the Azores-
Gibraltar Ridge at a mean depth of 4500 meters. Towing speeds
have been slowly increased and the most recent surveys were carried out at 7½ knots using the 22 km (12 mile) range, so that
7000 sq km (2000 sq miles) of the ocean floor can be surveyed per
day. In 1971 approximately 175000 sq km (50000 sq miles) was
surveyed in all. During this year an experimental long range fish
detection survey was also carried out on an inshore herring fishery
in the Western Isles of Scotland. It was found possible to detect
fish under favourable conditions out to a range of 18 km (10 miles
and to guide a purse seiner catching vessel on to the detected shoals.

149

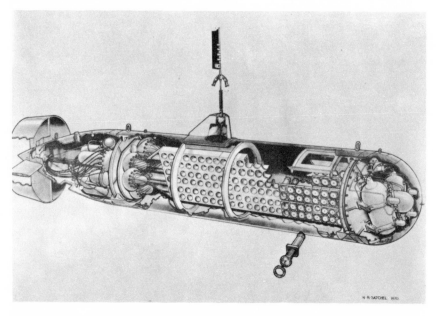

Fig. 7-5. A cut-away drawing of the sonar vehicle showing the rotatable array beam in the center section carrying 144 lead sinconate transducers.

Acoustic Positioning and Re-entry for Drilling

Floating drilling platforms are now operating in depths beyond 1000 feet and dynamic acoustic positioning systems are used. Measurement of headings, winds, waves and currents are recorded and interpreted by computers on board and the ship maintains position by the constant variations of speed/direction of its thrusters.

The drillship "Glomar Challenger" uses the Honeywell acoustic positioning system. This ship has now circumnavigated the world, obtaining core samples of the sediments and underlying rocks in all the deep ocean basins. The Scripps Institution managed the project under contract to the U.S. National Science Foundation and drilling took place in water depths of 2000 feet and bores of 5000 feet were made.

Further developments have taken place with the Edo Western Corporation who have developed a re-entry sonar. This system is

Fig. 7-6. A view of the inside of the vehicle showing the transducer array.

Fig. 7-7. A view of Palmer mountain ridge in the NE Atlantic with Freen Deep on the right. In the picture the ridge rises about 2,800 meters above the Deep which lies 5,200 meters beneath the ocean surface. The scene is 60 kilometers wide with a range of 14 kilometers, and has a vertical exaggeration of 2½.

lowered through the drill pipe and assists the drilling ship to re-enter a wellhead to continue a drilling programme.

Following the Mohole Project and the voyage of "Glomar Challenger", Shell commissioned in October 1971 the dynamic positioned drilling ship, to a design from the South Eastern Drilling Company, the "SEDCO 445". This has drilled in depths of 3000 feet.

Other drillships that have followed include the "Pelican" which uses a dynamic positioning system developed by the Institut Francais du Petrole (IFP), a research and development organization for the petroleum industry, in conjunction with Thompson CSF. And the "Havdrill", a sister ship which uses a system provided by Honeywell. Fig. 7-8.

Acoustic pipeline systems. Detection of pipeline leakage by acoustic means has not yet been developed, and at the moment there appears to be little work going on in this particular field. Of more interest for isolation of pipelines following failure is an

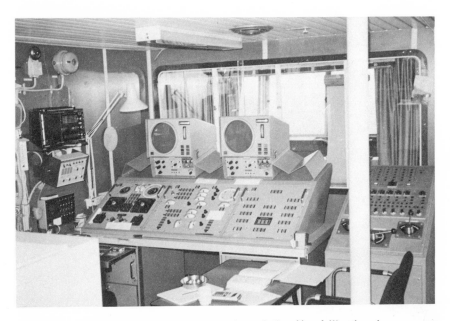

Fig. 7-8. Dynamic-position system aboard the Havdrill, showing operator's desk.

acoustically operated valve indication system. This enables undersea pipelines to be fitted with pipeline sectioning valves without the need for structures bringing the valve above water or, alternatively, divers to go down to manually close the valve.

The system is actuated by a coded sonar signal transmitted from either a ship or helicopter dunked sonar transmitter. Security is ensured by differences in coding of the valves on different pipelines in the same area. The only real disadvantage is that the battery life of the subsea valve operating system is only five years.

High-resolution sonar. Improvements are still being made to this type of equipment, mainly in the field of signal processing. A typical system manufactured by Marconi, which does not need to tow a body near the sea bottom, has a resolution of 0.3 meters at a range of 175 meters. This could be a useful tool for surveying and also remotely placing patterns of anchor blocks on the seabed without divers. As in most sonar display systems, a skilled operator is required to interpret the picture.

153

Diver-operated range navigation gear

The Havdrill, operated by Nordic Offshore Drilling Co.

This chapter has briefly and perhaps inadequately introduced the topic of communication and navigation to conclude the understanding of the range of engineering skills and activities involved in underwater engineering. It is perhaps in this area that the main advances in the future will be made, together with the development of subsea production techniques and remote controlled devices.